Assessment Book

Algebra 1

Explorations and Applications

This Assessment Book includes Program Objectives Record
Charts, Quizzes on groups of sections, Chapter Tests in
Forms A and B for each chapter, Cumulative Tests, and
Performance-Based Alternative Assessments for each chapter.
Answers are provided in a separate Answer Key.

McDougal Littell

A Houghton Mifflin Company

Evanston, Illinois

Boston   Dallas   Phoenix

Acknowledgment

The authors wish to thank Jane Pflughaupt, Mathematics Teacher, Pioneer High School, San Jose, California, for her valuable contributions to the *Assessment Book*.

ISBN: 0-395-76957-4

123456789 - PO - 99 98 97 96 95

CONTENTS

ALTERNATIVE ASSESSMENT   *Page*

NAME _____

PROGRAM OBJECTIVES RECORD CHART	Homework	Quiz	Test

CHAPTER 1 Using Algebra to Work with Data

		Homework	Quiz	Test
1.1	Use variables and inequalities. Analyze real-world data using histograms.			
1.2	Simplify an expression using the order of operations. Solve real-world problems.			
1.3	Find the mean, the median, and the mode(s) of a set of data. Analyze and compare real-world data.			
1.4	Find the absolute value of a number. Add and subtract integers.			
1.5	Multiply and divide integers. Use properties of addition and multiplication. Solve problems involving negative numbers.			
1.6	Write variable expressions. Solve real-world problems by using variables.			
1.7	Simplify variable expressions. Use the distributive property. Apply variable expressions in problem-solving situations.			
1.8	Organize data using spreadsheets and matrices. Work conveniently with large amounts of data.			

CHAPTER 2 Equations and Functions

		Homework	Quiz	Test
2.1	Write and solve one-step equations. Solve real-world problems including those involving distance, rate, and time.			
2.2	Write and solve two-step equations. Solve real-world problems and geometry problems using equations.			
2.3	Recognize and describe functions using tables and equations. Predict the outcome of a decision.			
2.4	Read and create coordinate graphs. Analyze biological and geographical data. Read and interpret maps.			
2.5	Graph equations. Visually represent functions.			
2.6	Use a graph to solve a problem. Analyze the costs and profits of running a business.			

NAME _____

CHAPTER 3   Graphing Linear Equations

	Homework	Quiz	Test
3.1 Find unit rates from words and graphs. Compare real-world rates.			
3.2 Recognize and describe direct variation. Explore relationships between real-world variables.			
3.3 Find the slope of a line. Analyze a real-world graph.			
3.4 Write linear equations in slope-intercept form. Analyze real-world problems represented by graphs.			
3.5 Find the *y*-intercept of a graph. Model real-world situations.			
3.6 Fit a line to data. Make predictions about real-world situations.			

CHAPTER 4   Solving Equations and Inequalities

	Homework	Quiz	Test
4.1 Model situations with tables and graphs. Compare options using tables and graphs.			
4.2 Solve one-step and two-step equations using reciprocals. Solve problems involving fractions.			
4.3 Solve multi-step equations. Use multi-step equations to solve real-world problems.			
4.4 Solve equations that involve more than one fraction. Use equations to solve real-world problems.			
4.5 Use inequalities to represent intervals on a graph. Use inequalities to represent a group of solutions to real-world problems.			
4.6 Solve inequalities. Tell when one quantity is greater than another.			

PROGRAM OBJECTIVES RECORD CHART

CHAPTER 5 **Connecting Algebra and Geometry**	Homework	Quiz	Test
5.1 Use ratios to compare quantities. Solve proportions. Estimate quantities that are difficult to count.			
5.2 Use scales in drawings and on graphs. Use scale factors to compare sizes of objects and drawings. Model large objects with scale drawings.			
5.3 Recognize and draw similar figures. Use similar figures to measure large distances.			
5.4 Find the ratios of the perimeters and areas of similar figures. Use scale drawings to find the perimeters and areas of large objects. Create accurate pictographs that represent statistics.			
5.5 Calculate theoretical and experimental probability. Predict the outcome of events.			
5.6 Find probabilities based on area. Predict events based on geometric probability.			

CHAPTER 6 **Working with Radicals**			
6.1 Identify right triangles using the Pythagorean theorem. Use the Pythagorean theorem to solve real-world problems.			
6.2 Find noninteger square roots. Estimate square roots to solve real-world problems.			
6.3 Multiply and divide radicals. Simplify radicals. Use skills with radicals to find values in real-world problems.			
6.4 Find products of monomials and binomials. Use products of monomials and binomials to find areas of complex figures.			
6.5 Recognize some special products of binomials. Find products of some variable expressions more quickly.			

PROGRAM OBJECTIVES RECORD CHART

	Homework	Quiz	Test
CHAPTER 7 **Systems of Equations and Inequalities**			
7.1 Write and graph equations in standard form. Solve problems with two variables.			
7.2 Solve systems of equations. Solve problems with two variables.			
7.3 Solve systems of equations by adding or subtracting. Solve problems with two variables.			
7.4 Write equivalent systems of equations. Solve equivalent systems of equations that model real-world situations.			
7.5 Graph linear inequalities. Visualize problems involving inequalities.			
7.6 Graph systems of inequalities. Graph systems of inequalities to model real-world situations when there are restrictions.			

CHAPTER 8   Quadratic Functions

	Homework	Quiz	Test
8.1 Analyze the shape of a graph. Decide whether a relationship is linear.			
8.2 Recognize characteristics of parabolas. Predict the shape of a parabola. Recognize side views of real objects.			
8.3 Use square roots and graphs to solve simple quadratic equations. Relate x-intercepts and solutions of quadratic equations. Solve simple motion problems involving acceleration.			
8.4 Use graphs to solve more complicated quadratic equations. Recognize quadratic equations with no solutions. Solve any quadratic equation and find heights of thrown objects.			
8.5 Use the quadratic formula. Solve quadratic equations without using a graph. Find more precise solutions to real-world problems.			
8.6 Find the discriminant of a quadratic equation. Find the number of solutions of a quadratic equation.			

PROGRAM OBJECTIVES RECORD CHART	Homework	Quiz	Test

CHAPTER 9   Exponential Functions

9.1	Show repeated multiplication with exponents. Evaluate expressions that involve powers. Find volumes of rectangular solids and spheres.			
9.2	Interpret and evaluate exponential functions. Model real-world situations with exponential functions. Make predictions using exponential functions.			
9.3	Use equations to model real-world decay situations. Make predictions about quantities that decrease.			
9.4	Evaluate powers with negative and zero exponents. Use negative exponents to express quotients. Find past values using negative exponents. Use formulas to find information.			
9.5	Write numbers in scientific notation.			
9.6	Multiply and divide expressions involving powers.			
9.7	Find the power of products and quotients. Find the power of a power. Use numbers written in scientific notation in formulas.			

CHAPTER 10   Polynomial Functions

10.1	Recognize and classify polynomials. Add and subtract polynomials. Use polynomials to model real-world situations.			
10.2	Use the *FOIL* method to multiply binomials. Multiply polynomials. Use polynomials to model lengths, areas, and volumes.			
10.3	Factor out a linear factor from a polynomial. Use the zero-product property. Use factoring to solve equations.			
10.4	Factor trinomials with positive quadratic terms and positive constant terms. Use factoring to explore the value of an investment.			
10.5	Factor quadratic polynomials with a negative constant term. Solve quadratic equations by factoring.			

PROGRAM OBJECTIVES RECORD CHART

	Homework	Quiz	Test
CHAPTER 11 **Rational Functions**			
11.1 Recognize and describe data that show inverse variation. Solve real-world problems involving inverse variation.			
11.2 Find weighted averages. Use weighted averages to solve real-world problems.			
11.3 Solve rational equations. Solve real-world problems involving rational equations.			
11.4 Solve a formula for one of its variables. Perform multiple calculations with the same formula.			
11.5 Multiply and divide rational expressions. Use multiplication and division of rational expressions to solve mathematical and real-world problems.			
11.6 Add and subtract rational expressions. Use addition and subtraction of rational expressions to solve real-world problems.			

CHAPTER 12   Discrete Mathematics

	Homework	Quiz	Test
12.1 Write and use algorithms. Use algorithms to understand and solve problems.			
12.2 Find the best path or tree for a graph.			
12.3 Recognize fairness in elections and divisions.			
12.4 Count all the ways groups of objects can be arranged.			
12.5 Count the number of possible pairs of objects in a group. Find the number of edges in a complete graph.			
12.6 Apply counting strategies to find probabilities. Find the probability of winning a random drawing.			

Test 1

QUIZ ON SECTIONS 1.1 THROUGH 1.3

DIRECTIONS: Write the answers in the spaces provided.

ANSWERS

Simplify each expression.

1. $24 \div 4 \cdot 3 + 7$ **2.** $36 - 2(8 + 6 \div 2)$ **3.** $\dfrac{4(5) - 8}{24}$

1. _____

2. _____

Evaluate each variable expression for $x = 5$.

4. $3x + \dfrac{4}{x}$ **5.** $x^2 + 3x - 4$ **6.** $\dfrac{x + 7}{9 - x}$

3. _____

Use the histogram for Exercises 7–10.

4. _____

7. How many students are represented by the histogram?

5. _____

8. How many students prefer 12 in. diameter or larger pizzas?

6. _____

9. Write two inequalities that describe the diameter of the pizza. Let $d =$ the diameter.

7. _____

10. Find the mean, the median, and the mode(s) of the data.

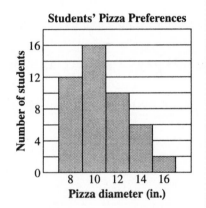

Students' Pizza Preferences

8. _____

9. _____

11. Writing List the steps in the order of operations.

10. *See question.* _____

11. *See question.* _____

NAME _____ DATE _____ SCORE _____

Test 2

DIRECTIONS: Write the answers in the spaces provided.

ANSWERS

Evaluate each variable expression when $m = -4$ and $w = 9$.

1. $15 - m + w$ 2. $6 - (m + w)$ 3. $w - 5 + m$

4. $|m|$ 5. $|-w| + m$ 6. $36 - |w - m|$

Simplify each expression.

7. $(-6)(-9)$ 8. $(-8)(4)$ 9. $-28 \div (-4)$

For Questions 10–12, evaluate each variable expression when $x = -3$ and $y = 9$.

10. $x - y - 5$ 11. $2x \cdot \dfrac{5}{6} \cdot y$ 12. $-3y + x + 30$

13. **Writing** Explain the meaning of the term *absolute value*. Give examples to illustrate your explanation.

1. _____

2. _____

3. _____

4. _____

5. _____

6. _____

7. _____

8. _____

9. _____

10. _____

11. _____

12. _____

13. *See question.* _____

Test 3

QUIZ ON SECTIONS 1.6 THROUGH 1.8

DIRECTIONS: Write the answers in the spaces provided.

Write a variable expression.

1. the number of days in w weeks

2. the number of dollars in a bank account that contained d dollars before a deposit of $175

Simplify each variable expression.

3. $7(x - 4)$ **4.** $5\left(x - \frac{3}{5}\right) + 2x$ **5.** $8x^2 - 9y - 2x + 5y$

For Questions 6 and 7, use the spreadsheet below.

	A	B	C
1		Regular price	Sale price
2	Shirts	$22.50	$17.95
3	Shorts	$19.95	$15.00
4	Hats	$7.50	$5.99

6. What formula can you enter to find the total sale price of 3 shirts and 2 hats?

7. What is the total sale price of 3 shirts and 2 hats?

For Questions 8 and 9, use matrices *P*, *Q*, and *R* below.

$$P = \begin{bmatrix} 4 & -6 \\ -3 & 7 \end{bmatrix} \qquad Q = \begin{bmatrix} -5 & 2 & 11 \\ 3 & -7 & 9 \\ 8 & 0 & -1 \end{bmatrix} \qquad R = \begin{bmatrix} -5 & 2 \\ 6 & 3 \end{bmatrix}$$

8. Give the dimensions of each matrix.

9. Which matrices can be added? Add them.

10. Writing How are matrices and spreadsheets alike? How are they different?

ANSWERS

1. _____

2. _____

3. _____

4. _____

5. _____

6. _____

7. _____

8. *See question.*

9. *See question.*

10. *See question.*

Test 4

TEST ON CHAPTER 1 (FORM A)

DIRECTIONS: Write the answers in the spaces provided.

For Questions 1–3, evaluate each variable expression for $a = 4$, $b = 6$, and $c = 8$.

1. $3a + 2c$

2. $6a - 3b$

3. $\dfrac{c^2 - b^2}{a}$

4. Find the mean, the median, and the mode(s) for the ages below.
20, 16, 14, 15, 19, 16, 15, 17, 18, 13

Simplify each expression.

5. $8 - 13$

6. $(-5) - (-2)$

7. $(-8) + (-7)$

8. $(-3)(-9)$

9. $-24 \div 8$

10. $|-6| + |2|$

11. $|-5 + 2|$

12. $-5 + |4|$

13. $(-6)(7)$

Evaluate each variable expression for $x = -4$ and $y = 6$.

14. $3x + 5y + x - 2y$

15. $2x^2 + 4xy - y^2 - 2x^2$

Simplify each variable expression.

16. $3x - 8 + 12x + 19$

17. $7 + 4x + 3y - x + 8y$

18. $-6(5x - y) + 4(2x - 3y)$

19. $5x^2 - 3x - 17x^2 + 12x$

For Questions 20–22, use matrices A, B, and C.

$A = \begin{bmatrix} 2 & 6 \\ 5 & 7 \\ 4 & 3 \end{bmatrix}$ $B = \begin{bmatrix} 8 & -6 \\ 2 & 1 \end{bmatrix}$ $C = \begin{bmatrix} 3 & -1 \\ 0 & 2 \\ -5 & 4 \end{bmatrix}$

20. What are the dimensions of each matrix?

21. Which two matrices can be added together?

22. Add the two matrices you listed in Question 21.

ANSWERS

1. _____

2. _____

3. _____

4. *See question.*

5. _____

6. _____

7. _____

8. _____

9. _____

10. _____

11. _____

12. _____

13. _____

14. _____

15. _____

16. _____

17. _____

18. _____

19. _____

20. *See question.*

21. _____

22. *See question.*

Test 4

(CONTINUED)

DIRECTIONS: Write the answers in the spaces provided.

23. There are 12 in. in one foot. Write a variable expression for the number of inches in *f* feet.

24. Melissa is selling orange juice and doughnuts at a school fundraiser. Orange juice sells for $1.25 each and doughnuts sell for $.60 each. Write a variable expression that represents Melissa's income if each of her friends buys one orange juice and one doughnut. Let *n* = the number of her friends buying food. Then find Melissa's income if she sells one orange juice and one doughnut to each of 11 friends.

25. **Writing** Sandeep recorded the time it took each member of the track team to run a mile. He entered the data in a spreadsheet with minutes in column A and seconds in column B. Each runner's time will be given in seconds in column C. Explain what the spreadsheet program must do to find the values for column C.

ANSWERS

23. _____

24. _____

25. *See question.*

26. _____

27. _____

28. *See question.*

For Questions 26–28, use the histogram at the right.

26. For how many hours is the parking lot full?

27. How many cars does it take to fill the parking lot?

28. **Open-ended Problem** Describe a real-life situation that the histogram data might represent.

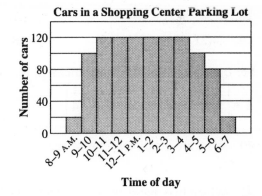

Cars in a Shopping Center Parking Lot

Test 5 ···

TEST ON CHAPTER 1 (FORM B)

DIRECTIONS: Write the answers in the spaces provided.

For Questions 1–3, evaluate each variable expression for _m_ = 5, _r_ = 8, and _t_ = 10.

1. $4r - 2t$

2. $3r + m$

3. $\dfrac{r^2 - 3r}{t}$

4. Find the mean, the median, and the mode(s) for the ages below.
25, 21, 19, 20, 24, 21, 20, 22, 23, 18

Simplify each expression.

5. $9 - 16$

6. $(-3) - (-7)$

7. $(-5) + (-9)$

8. $(-4)(-8)$

9. $-42 \div 7$

10. $|-8| + |3|$

11. $|-7 + 3|$

12. $-8 + |5|$

13. $(-9)(6)$

Evaluate each variable expression for _x_ = –6 and _y_ = 4.

14. $2x + 6y - x + 3y$

15. $y^2 + 5xy - 2x^2 - y^2$

Simplify each variable expression.

16. $6x - 9 + 3x + 14$

17. $9 - 5x + 13y - x - 10y$

18. $-4(2x - 3y) + 3(5x - y)$

19. $3x^2 + 5x - 12x^2 - 3x$

For Questions 20–22, use matrices _A_, _B_, and _C_.

$$A = \begin{bmatrix} 2 & 5 & -4 \\ 5 & 9 & 0 \\ 4 & 2 & -6 \end{bmatrix} \qquad B = \begin{bmatrix} -5 & 1 \\ 8 & -4 \\ 2 & -2 \end{bmatrix} \qquad C = \begin{bmatrix} 3 & -2 \\ 0 & 3 \\ -5 & 1 \end{bmatrix}$$

20. What are the dimensions of each matrix?

21. Which two matrices can be added together?

22. Add the two matrices you listed in Question 21.

ANSWERS

1. _____

2. _____

3. _____

4. _See question._ ___

5. _____

6. _____

7. _____

8. _____

9. _____

10. _____

11. _____

12. _____

13. _____

14. _____

15. _____

16. _____

17. _____

18. _____

19. _____

20. _See question._ __

21. _____

22. _See question._ __

Test 5

(CONTINUED)

DIRECTIONS: Write the answers in the spaces provided.

ANSWERS

23. There are 3 ft in one yard. Write a variable expression for the number of feet in *y* yards.

24. Stanley is selling helium-filled balloons at a school fundraiser. Metallic balloons sell for $2.75 each and regular balloons sell for $1.25 each. Write a variable expression that represents Stanley's income if each of his friends buys one balloon of each type. Let *n* = the number of his friends buying balloons. Then find Stanley's income if he sells one metallic balloon and one regular balloon to each of 9 friends.

25. **Writing** Kenyatta recorded the height of each member of the dance team in feet and inches. She entered the data in a spreadsheet with feet in column A and inches in column B. Explain what the spreadsheet program must do to find each dance team member's total height in inches for column C.

23. _____

24. _____

25. *See question.* _____

26. _____

27. _____

28. *See question.* _____

For Questions 26–28, use the histogram at the right.

26. During which two days do students work the most hours?

27. Find the mean number of hours that the students work per day.

28. **Open–ended Problem** Describe a real-life situation that the histogram data might represent.

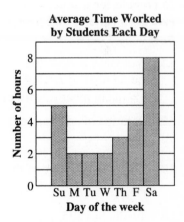

Average Time Worked by Students Each Day

Number of hours (vertical axis: 0, 2, 4, 6, 8)

Day of the week (horizontal axis: Su M Tu W Th F Sa)

Test 6

QUIZ ON SECTIONS 2.1 THROUGH 2.3

DIRECTIONS: Write the answers in the spaces provided.

ANSWERS

For Questions 1–9, solve each equation.

1. $-4 = 20x$ **2.** $-m - 9 = 4$ **3.** $-8 = 12 + r$

4. $36 - 9p = 0$ **5.** $-z = 14$ **6.** $-24 = 3d$

7. $25 = 19 + \frac{m}{3}$ **8.** $\frac{v}{9} = -7$ **9.** $\frac{t}{6} + 5 = -3$

10. Open-ended Problem Write an equation using at least three of the numbers in the following list: 2, 5, 10, 15, 20, 25, 50, 100. Then write a word problem that describes a real-life situation that can be solved using your equation. Finally, solve your equation and state the answer to your problem.

To prepare for a school fundraiser, Juanita pays $5.00 for 100 cups. She also buys cans of fruit juice for $.50 each. Use this information for Questions 11–14.

11. Write an equation that describes Juanita's total cost as a function of the number of cans of fruit juice she buys. Let C = the total cost and n = the number of cans of fruit juice purchased.

12. Describe the domain of the function.

13. Describe the range of the function.

14. Juanita has $35 to spend on cups and fruit juice for the fundraiser. If she buys 100 cups, how many cans of fruit juice can she buy?

1. _____

2. _____

3. _____

4. _____

5. _____

6. _____

7. _____

8. _____

9. _____

10. *See question.*

11. _____

12. *See question.*

13. *See question.*

14. _____

NAME _____ DATE _____ SCORE _____

Test 7 ···

QUIZ ON SECTIONS 2.4 THROUGH 2.6

DIRECTIONS: Write the answers in the spaces provided.

Graph each point in the coordinate plane at the right. Label each point with its letter. Name the quadrant (if any) in which the point lies.

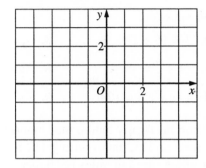

1. $Z(-2, 3)$ **2.** $Y(0, -2)$

3. $X(3, -3)$ **4.** $U(1, 2)$

5. $T(-4, -3)$ **6.** $S(-3, 2)$

Graph the equation and the ordered pair. Tell whether the ordered pair is a solution of the equation. Write *Yes* or *No*.

7. $y = 2x - 1$; $(-1, -1)$ **8.** $y = -\frac{1}{3}x + 2$; $(3, 1)$

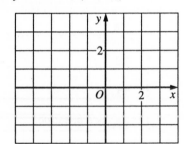   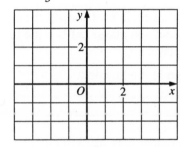

A neighborhood band pays $145 for decorations and refreshments for a local dance. The band sells tickets for $5 each. Use this information for Questions 9 and 10.

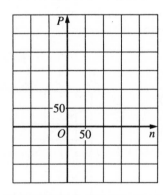

9. Write and graph an equation for their profit P from selling n tickets.

10. How many tickets must the band sell in order to make a profit of $250?

11. Writing Explain how you can show that an ordered pair is a solution of an equation.

ANSWERS

1. _____

2. _____

3. _____

4. _____

5. _____

6. _____

7. _____

8. _____

9. _____

10. _____

11. *See question.* _____

Test 8 ···

TEST ON CHAPTER 2   (FORM A)

DIRECTIONS: Write the answers in the spaces provided.

ANSWERS

For Questions 1–9, solve each equation.

1. $5t + 125 = 600$ **2.** $3 = 15 + 12x$ **3.** $-z = 51$

4. $\dfrac{x}{2} - 8 = 2$ **5.** $-1 + \dfrac{n}{6} = -3$ **6.** $13 - b = 18$

7. $-p = -43$ **8.** $-4r + 2.6 = 10.6$ **9.** $-q - 10 = -20$

10. Last summer, Jack Smith and his family toured the United States for 2 weeks. They traveled a total of 5152 miles during 112 hours of driving time. Find their average speed in miles per hour.

Felicia Johnson paid $125 to join a tennis club. She pays an additional $5 every time she uses one of the club's tennis courts. Use this information for Questions 11–13.

11. Write an equation that describes Felicia's total cost for playing tennis as a function of the number of times she plays. Let C = the total cost and n = the number of times she plays.

12. Describe the domain and range of the function.

13. Felicia does not want to spend more than $275 to play tennis during the summer. What is the maximum number of times that she can play tennis on the club's courts for this amount?

The graph at the right shows the temperature in degrees Fahrenheit in a Minnesota city during a 24-hour period last winter. Use this information for Questions 14 and 15.

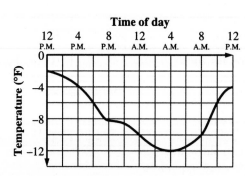

Time of day

14. What was the highest temperature during this 24-hour period?

15. At what hour was the temperature about $-12°F$?

1. _____

2. _____

3. _____

4. _____

5. _____

6. _____

7. _____

8. _____

9. _____

10. _____

11. _____

12. *See question.*

13. _____

14. _____

15. _____

Test 8

(CONTINUED)

DIRECTIONS: Write the answers in the spaces provided.

ANSWERS

Graph each point in the coordinate plane at the right. Label each point with its letter. Name the quadrant (if any) in which the point lies.

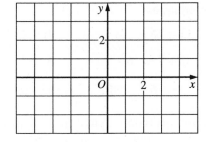

16. $P(0, 3)$ **17.** $D(-2, 1)$

18. $B(-2, -1)$ **19.** $N(1, -1)$

20. $T(-3, 2)$ **21.** $E(3, 3)$

For Questions 22 and 23, graph each equation.

22. $y = -3x$ **23.** $y = 2x - 3$

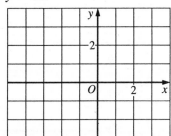
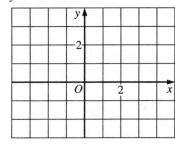

16. _____

17. _____

18. _____

19. _____

20. _____

21. _____

22. *See question.*

23. *See question.*

24. *See question.*

25. _____

26. _____

27. *See question.*

24. Open-ended Problem Describe the steps you took to graph the equation in Question 23. Describe another way to graph the equation.

The O'Briens are planning an anniversary party. They are going to rent a banquet hall for $350, and the caterer will charge $25 per person. Use this information for Questions 25–27.

25. Write and graph an equation describing their cost C for the hall and catering as a function of the number of people n.

26. The O'Briens have budgeted $1800 for the banquet hall and catering. What is the maximum number of people who can attend the party for this amount?

27. Writing What method did you use to solve Question 26? Describe one other way to solve the problem.

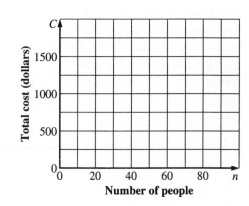

Test 9 ·······································

TEST ON CHAPTER 2   **(FORM B)**

DIRECTIONS: Write the answers in the spaces provided.

ANSWERS

For Questions 1–9, solve each equation.

1. $25v + 125 = 600$ **2.** $33 = 15 + 12x$ **3.** $-z = 59$

4. $\frac{x}{2} + 8 = 2$ **5.** $-1 + \frac{n}{6} = -6$ **6.** $13 - b = -18$

7. $-p = -52$ **8.** $-4r + 6.7 = 10.7$ **9.** $-q - 50 = -20$

10. Last summer, Rico Flores and his family toured the United States for 3 weeks. They traveled a total of 7056 miles during 168 hours of driving time. Find their average speed in miles per hour.

Ranil Punjab paid $150 to join a racquetball club. He pays an additional $15 every time he uses one of the club's racquetball courts. Use this information for Questions 11–13.

11. Write an equation that describes Ranil's total cost for playing racquetball as a function of the number of times he plays. Let C = the total cost and n = the number of times he plays.

12. Describe the domain and range of the function.

13. Ranil does not want to spend more than $525 to play racquetball during the summer. What is the maximum number of times that he can play racquetball on the club's courts for this amount?

The graph at the right shows the temperature in degrees Fahrenheit in a Michigan city during a 24-hour period last winter. Use this information for Questions 14 and 15.

14. What was the lowest temperature during this 24-hour period?

15. At what hour was the temperature about −12°F?

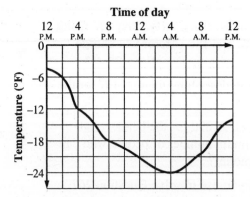

1. _____

2. _____

3. _____

4. _____

5. _____

6. _____

7. _____

8. _____

9. _____

10. _____

11. _____

12. *See question.*

13. _____

14. _____

15. _____

Test 9

DIRECTIONS: Write the answers in the spaces provided.

Graph each point in the coordinate plane at the right. Label each point with its letter. Name the quadrant (if any) in which the point lies.

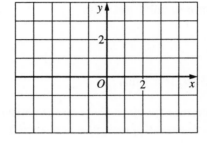

16. $P(-1, 3)$ **17.** $D(3, 1)$

18. $A(-3, 0)$ **19.** $N(-4, -1)$

20. $T(3, -2)$ **21.** $E(2, 2)$

For Questions 22 and 23, graph each equation.

22. $y = -2x$

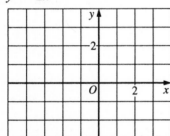

23. $y = 3x - 2$

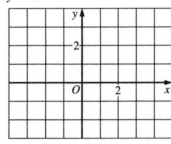

ANSWERS

16. _____

17. _____

18. _____

19. _____

20. _____

21. _____

22. *See question.*

23. *See question.*

24. *See question.*

25. _____

26. _____

27. *See question.*

24. Open-ended Problem Describe the steps you took to graph the equation in Question 23. Describe another way to graph the equation.

The Daniels are planning a graduation party for their son. They are hiring a band to play for $500 and are figuring $15 per person for food. Use this information for Questions 25–27.

25. Write and graph an equation describing their cost C for the band and food as a function of the number of people n.

26. The Daniels have budgeted $1700 for the band and food. What is the maximum number of people who can attend the party for this amount?

27. Writing What method did you use to solve Question 26? Describe one other way to solve the problem.

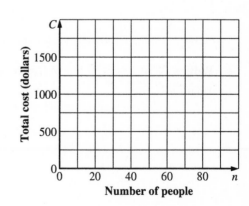

Test 10

QUIZ ON SECTIONS 3.1 THROUGH 3.2

DIRECTIONS: Write the answers in the spaces provided.

Express each rate as a unit rate in miles per hour.

1. 2000 mi in 40 h

2. 0.5 mi in 3 min

3. 1 foot per second (Hint: 5280 ft = 1 mi)

For Questions 4 and 5, tell whether the data show direct variation. Write *yes* or *no*. If they do, give the constant of variation and write an equation that describes the situation.

4.

Person's age (years)	Person's height (inches)
10	60
11	61
12	63
13	64
14	67

5.

Donuts sold	Money collected (dollars)
5	2
10	4
15	6
20	8
25	10

6. **Writing** Write a definition of *direct variation*. What is the *constant of variation*?

ANSWERS

1. _____

2. _____

3. _____

4. _____

5. _____

6. *See question.*
_____

NAME _____ DATE _____ SCORE _____

Test 11

DIRECTIONS: Write the answers in the spaces provided.

ANSWERS

Find the slope of the line through each pair of points.

1. $(5, 6), (-1, 3)$ **2.** $(-8, -2), (-3, -4)$ **3.** $(0, 6), (6, 0)$

The water in a fish tank is evaporating at a constant rate. The graph at the right shows the depth of the water as a function of time.

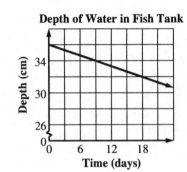

Depth of Water in Fish Tank

4. Write an equation for the line in slope-intercept form.

5. What information do the slope and vertical intercept tell you about the situation?

1. _____

2. _____

3. _____

4. _____

5. *See question.* _____

6. _____

7. _____

8. *See question.* _____

For Questions 6 and 7, graph each equation. Give the slope and *y*-intercept of the graph.

6. $y = 2x - 3$

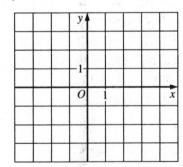

7. $y = -\frac{5}{3}x + 4$

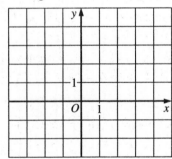

8. Writing Explain the difference between positive slope and negative slope. Include sketches of lines showing each type of slope.

Test 12

QUIZ ON SECTIONS 3.5 THROUGH 3.6

DIRECTIONS: Write the answers in the spaces provided.

ANSWERS

For Questions 1–6, find an equation of the line through the given points.

1. (5, 6), (3, –3) **2.** (8, 2), (–2, 5) **3.** (5, 3), (5, –3)

4. (7, 2), (–4, 2) **5.** (–1, 1), (–4, –14) **6.** (3, –6), (–5, 2)

7. Graph the line that passes through the points (–4, 0) and (0, 3).

8. Write an equation of the line you graphed for Question 7.

9. Write an equation of the line that passes through the point (3, 0) and has the same slope as the line in Question 7.

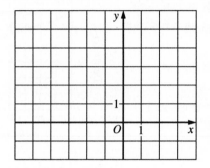

The table below lists the average test scores for several tests in one teacher's class. Use the table for Questions 10–12.

Test	1	2	3	4	5	6
Average score	60	50	75	80	85	80

10. Make a scatter plot of the data.

11. Draw a line of fit for the data. Find an equation for your line.

12. Predict the average score on the next test given in this class.

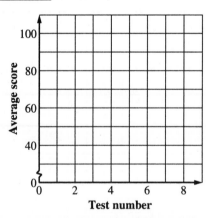

1. _____

2. _____

3. _____

4. _____

5. _____

6. _____

7. *See question.* _____

8. _____

9. _____

10. *See question.* _____

11. _____

12. _____

Test 13

DIRECTIONS: Write the answers in the spaces provided.

ANSWERS

Complete each equation.

1. $\dfrac{60 \text{ mi}}{1 \text{ h}} \cdot \dfrac{24 \text{ h}}{1 \text{ day}} = \underline{\ \ ?\ \ }$

2. $\dfrac{12 \text{ in.}}{1 \text{ ft}} \cdot \underline{\ \ ?\ \ } = \dfrac{36 \text{ in.}}{1 \text{ yd}}$

For Questions 3 and 4, express each rate in the given units.

3. $\dfrac{\$12.60}{7 \text{ yd}} = \underline{\ \ ?\ \ }$ cents per inch

4. $\dfrac{3.6 \text{ mi}}{3 \text{ min}} = \underline{\ \ ?\ \ }$ mi/h

5. Which graph is a model of the direct variation $y = 5x$?

A. 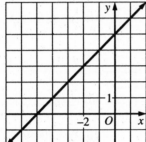   B.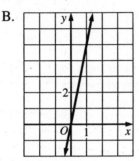

6. The amount of gasoline used by a car varies directly with the number of miles traveled. It is estimated that a certain new car averages 1 gal of gas for every 32 mi traveled. About how many gallons of gas will be used if this car is driven 2500 mi?

7. Decide whether the data in the table at the right show direct variation. If they do, give the constant of variation and write an equation in the form $y = kx$.

Length	Area
2	6
3	9
5	15
11	33

8. **Writing** Explain the difference between a horizontal line and a vertical line in terms of slope.

Find the slope of the line passing through each pair of points.

9. (12, 5) and (6, 10)

10. (−8, 2) and (6, −2)

1. _____

2. _____

3. _____

4. _____

5. _____

6. _____

7. _____

8. *See question.* _____

9. _____

10. _____

Test 13

(CONTINUED)

DIRECTIONS: Write the answers in the spaces provided.

For Questions 11 and 12, graph each equation.

11. $y = x - 1$

12. $y = -\frac{1}{3}x + 2$

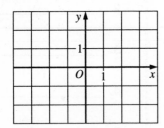

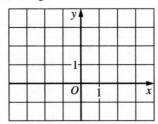

13. Find an equation of a line with slope 2 and through $(5, -3)$.

14. Writing Write an equation in slope-intercept form for the graph shown at the right. What are the slope and vertical intercept of the graph, and what do they tell you about the graph?

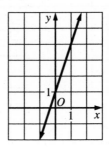

11. _See question._

12. _See question._

13. _____

14. _See question._

15. _____

16. _____

17. _____

18. _See question._

19. _See question._

20. _____

Find an equation of the line through each pair of points.

15. $(4, -2), (2, 3)$ **16.** $(-3, -1), (6, 2)$ **17.** $(5, 8), (-1, 8)$

For Questions 18–20, use the data in the table at the right.

18. Draw a scatter plot of the data. Then draw a line of fit on your scatter plot.

19. Write an equation for your line of fit.

20. Use your equation to predict the shoe size of a 74 in. tall man.

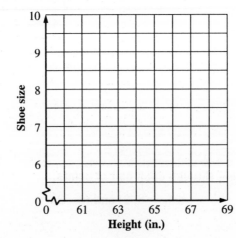

Height of a man (in.)	Shoe size
61	7.5
63	8.5
65	9
67	9
69	10

Test 14

TEST ON CHAPTER 3   (FORM B)

DIRECTIONS: Write the answers in the spaces provided.

ANSWERS

Complete each equation.

1. $\dfrac{50 \text{ words}}{1 \text{ min}} \cdot \dfrac{60 \text{ min}}{1 \text{ h}} = \underline{\ ?\ }$

2. $\dfrac{24 \text{ h}}{1 \text{ day}} \cdot \underline{\ ?\ } = \dfrac{168 \text{ h}}{1 \text{ wk}}$

For Questions 3 and 4, express each rate in the given units.

3. $\dfrac{\$10.80}{9 \text{ ft}} = \underline{\ ?\ }$ cents per inch

4. $\dfrac{2.5 \text{ mi}}{4 \text{ min}} = \underline{\ ?\ }$ mi/h

5. Which graph is a model of the direct variation $y = 3x$?

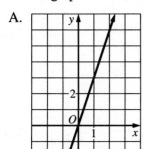   **A.**

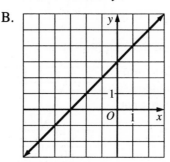   **B.**

6. The amount of gasoline used by a car varies directly with the number of miles traveled. It is estimated that a certain new car averages 1 gal of gas for every 26 mi traveled. About how many gallons of gas will be used if this car is driven 3000 mi?

7. Decide whether the data in the table at the right show direct variation. If they do, give the constant of variation and write an equation in the form $y = kx$.

Base	Area
10	20
12	24
15	30
19	38

8. Writing Explain what a scatter plot is and why it is useful.

Find the slope of the line passing through each pair of points.

9. (2, 5) and (6, 0)

10. (−8, 4) and (6, −4)

1. _____

2. _____

3. _____

4. _____

5. _____

6. _____

7. _____

8. *See question.*
_____

9. _____

10. _____

Test 14

(CONTINUED)

DIRECTIONS: Write the answers in the spaces provided.

For Questions 11 and 12, graph each equation.

11. $y = x - 3$

12. $y = -\frac{1}{2}x + 2$

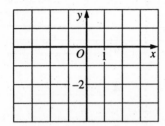

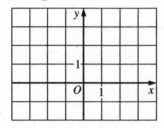

13. Find an equation of a line with slope -2 and through $(4, -1)$.

14. Writing Write an equation in slope-intercept form for the graph shown at the right. What are the slope and vertical intercept of the graph, and what do they tell you about the graph?

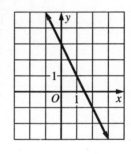

Find an equation of the line through each pair of points.

15. $(5, -2), (2, 4)$ **16.** $(5, 3), (-1, 3)$ **17.** $(4, 3), (-6, -2)$

For Questions 18–20, use the data in the table at the right.

18. Draw a scatter plot of the data. Then draw a line of fit on your scatter plot.

19. Write an equation for your line of fit.

20. Use your equation to predict the weight of the winning pumpkin in the tenth year.

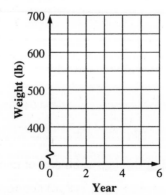

Year	Weight of winning pumpkin (lb)
1	550
2	500
3	550
4	650
5	600

ANSWERS

11. *See question.* _____

12. *See question.* _____

13. _____

14. *See question.* _____

15. _____

16. _____

17. _____

18. *See question.* _____

19. *See question.* _____

20. _____

20

Test 15

DIRECTIONS: Write the answers in the spaces provided.

A movie theater offers two types of movie club memberships to patrons. The Premium Membership has an annual fee of $30 for which a member receives free popcorn all year long when paying the regular admission price of $5.00 for each movie attended. The Elite Membership has an annual fee of $200 for which a member receives admission to an unlimited number of movies during the year as well as free popcorn.

1. Complete the table at the right to model this situation.

2. For what number of movies do the two memberships cost the same?

Number of movies	Cost of Premium Membership	Cost of Elite Membership
10		$200
15		$200
20		$200
25		$200
30		$200
35		$200
40		$200

ANSWERS

1. _See question._
2. _____
3. _____
4. _____
5. _____
6. _____
7. _____
8. _____
9. _____
10. _____
11. _See question._

For Questions 3–10, solve each equation. If the equation is an _identity_ or there is _no solution_, say so.

3. $\frac{3}{4}c = 12$

4. $-\frac{5}{6}w + 3 = 18$

5. $\frac{4x}{5} = -16$

6. $-\frac{3}{8}x + 2 = -10$

7. $8y + 56 = 4y$

8. $-3p - 5 = 8p + 17$

9. $7x + 15 = x + 3(5 + 2x)$

10. $3(2a - 5) = 2(3a + 1)$

11. **Writing** Explain how a reciprocal and the multiplicative identity can be used to solve an equation such as $\frac{2}{5}x = 18$.

Test 16

QUIZ ON SECTIONS 4.4 THROUGH 4.6

DIRECTIONS: Write the answers in the spaces provided.

ANSWERS

Solve each equation.

1. $\dfrac{x}{4} = \dfrac{5}{3}$

2. $\dfrac{11}{6}y - \dfrac{5}{6} = \dfrac{1}{12}$

3. $\dfrac{3}{4} - \dfrac{z}{5} = \dfrac{1}{20}$

4. $0.5x + 0.3 = 1.8$

5. $3p - 0.8 = 4.9$

6. $4.8 - 0.15m = 0.18$

For Questions 7–11, solve each inequality. Graph each solution on a number line.

7. $x + 2 < 6$

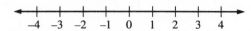

8. $\dfrac{p}{-7} \geq 2$

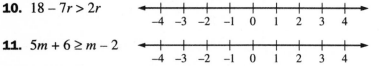

9. $5 - 3x < 14$

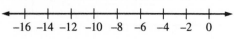

10. $18 - 7r > 2r$

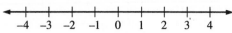

11. $5m + 6 \geq m - 2$

12. Maria is the hostess and manager at a restaurant. She receives a yearly salary of $18,500, plus 4% of each day's receipts at the restaurant. Last year Maria's total pay was more than $36,000. Write and solve an inequality based on this situation.

13. **Writing** Explain how you choose a power of ten to multiply both sides of an equation by when solving an equation that involves decimals.

1. _____

2. _____

3. _____

4. _____

5. _____

6. _____

7. _____

8. _____

9. _____

10. _____

11. _____

12. *See question.*
_____

13. *See question.*
_____

NAME _____ DATE _____ SCORE _____

Test 17

DIRECTIONS: Write the answers in the spaces provided.

An ice skating rink charges $10 for two hours of skating. Renting skates for the two hours costs $8. Suppose you can buy a pair of ice skates for $110.

1. Make a graph that models this situation.

2. When will it begin to cost more if you rent skates each time than if you buy skates?

3. Write an equation to model the situation.

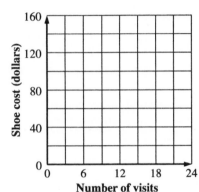

For Questions 4–6, solve each equation.

4. $\frac{3}{4}x = 15$

5. $\frac{5n}{6} - 10 = -25$

6. $9w - 5 = 5w + 3$

7. Which equation below is an identity? Which has no solutions?
   A. $10(x + 3) + 8 = 18x + 30$
   B. $16n - 20 = 4(5n + 1)$
   C. $12(c + 3) - 30 = 12c + 36$
   D. $4(6a + 3) = 6(4a + 2)$

For Questions 8–11, solve each equation. If an equation is an *identity* or if there is *no solution*, say so.

8. $-25x + 12 = -15x - 8$

9. $-3w - 10 + 5w = 2(w - 5)$

10. $\frac{1}{3}(6x - 15) = 2x + 7$

11. $5m - 3 = \frac{2}{3}(9m - 6) + 3$

12. A season ticket for a rinkside seat at a hockey arena costs $750 for 25 games. The cost of buying a ticket for the same seat for one game is $40. Write and solve an equation to find out how many games a fan will have to attend to recover the cost of a season ticket.

ANSWERS

1. *See question.* _____

2. _____

3. _____

4. _____

5. _____

6. _____

7. _____

8. _____

9. _____

10. _____

11. _____

12. *See question.* _____

Test 17

(CONTINUED)

DIRECTIONS: Write the answers in the spaces provided.

ANSWERS

13. What is the smallest number that you can multiply by to eliminate the fractions in the equation $\frac{1}{2} + \frac{9}{10}m = \frac{4}{15} - \frac{1}{6}m$?

13. _____

14. _____

For Questions 14–17, solve each equation.

14. $\frac{3}{4}a + 1 = \frac{2}{3}a - 5$ 15. $\frac{5}{6}r - \frac{1}{2} = \frac{2}{15}r + \frac{3}{5}$

15. _____

16. $0.8p + 5 = 11.3 - p$ 17. $6.496 - 2.4d = 1.66d$

16. _____

18. If $s > c$, which of the following inequalities is *not* true?

A. $-4s > -4c$ B. $\frac{s}{5} > \frac{c}{5}$ C. $s - (-3) > c - (-3)$

17. _____

Solve each inequality. Graph each solution on a number line.

18. _____

19. $8 + 5x < -2$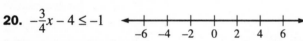

20. $-\frac{3}{4}x - 4 \le -1$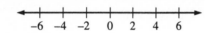

19. _____

20. _____

The graph at the right shows the number of cars in a parking lot as a function of time of day. Use the graph for Questions 21 and 22.

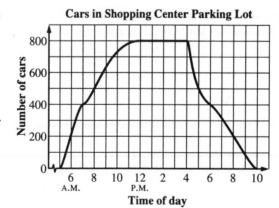

Cars in Shopping Center Parking Lot

21. _See question._

22. _____

21. At what time is the parking lot less than half full? Write your answer as two inequalities.

23. _See question._

22. At what time does the parking lot have at least 500 cars parked in it? Write your answer as an inequality.

23. **Open-ended Problem** Make up an inequality that changes the direction of the inequality sign when the inequality is being solved. Show the correct steps for finding the solution of your inequality.

NAME _____ DATE _____ SCORE _____

Test 18

DIRECTIONS: Write the answers in the spaces provided.

ANSWERS

A bowling alley charges $4 for each game bowled. Renting bowling shoes costs $0.75 for each visit to the bowling alley. Suppose you can buy a pair of bowling shoes for $41.

1. Make a graph that models this situation.

2. When will it begin to cost more if you rent shoes each time than if you buy shoes?

3. Write an equation to model the situation.

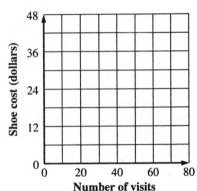

1. *See question.* _____

2. _____

3. _____

4. _____

5. _____

6. _____

For Questions 4–6, solve each equation.

4. $\frac{2}{3}x = 16$

5. $\frac{3w}{8} - 6 = -12$

6. $7y - 8 = 3y + 4$

7. Which equation is an identity? Which equation has no solutions?

   A. $15(p + 3) + 12 = 27p + 45$ B. $6(6t + 3) = 9(4t + 2)$

   C. $24m - 30 = 2(15m + 3)$ D. $18(w + 3) - 45 = 18w + 54$

7. _____

8. _____

9. _____

10. _____

11. _____

12. *See question.* _____

For Questions 8–11, solve each equation. If an equation is an *identity* or if there is *no solution*, say so.

8. $-40w + 16 = -6w - 1$

9. $7k - 4 - 3k = 4k + 7$

10. $\frac{1}{4}(8x - 12) = 3x + 5$

11. $\frac{5}{6}(12p + 18) - 3 = 10p + 12$

12. A season ticket for a first baseline seat at a baseball stadium costs $400 for 42 games. The cost of buying a ticket for the same seat for one game is $12. Write and solve an equation to find out how many games a fan will have to attend to recover the cost of a season ticket.

Test 18 ·······················

(CONTINUED)

DIRECTIONS: Write the answers in the spaces provided.

ANSWERS

13. What is the smallest number that you can multiply by to eliminate the fractions in the equation $\frac{5}{8}w - \frac{3}{4} = \frac{1}{12}w + \frac{5}{6}$?

13. _____

14. _____

For Questions 14–17, solve each equation.

14. $\frac{2}{5}k - 3 = \frac{3}{4}k - 2$ **15.** $\frac{5}{16}w + \frac{1}{8} = \frac{5}{12}w + \frac{3}{4}$

15. _____

16. $3.4m + 8 = 4.9 - 5.2m$ **17.** $0.8h + 0.26 = 7.1 - 0.72h$

16. _____

18. If $d < f$, which of the following inequalities is true?

   A. $d - 8 > f - 8$ B. $-3 + d > -3 + f$ C. $2d > 2f$

17. _____

Solve each inequality. Graph each solution on a number line.

18. _____

19. $8 + 5x < -7$

19. _____

20. $-\frac{2}{3}x - 5 \le -7$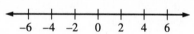

20. _____

The graph at the right shows the number of calories burned per hour during a workout. Use the graph for Questions 21 and 22.

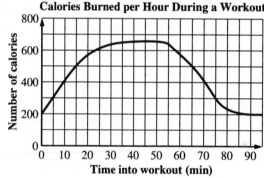

Calories Burned per Hour During a Workout

21. *See question.*

21. At what time is the person burning less than 300 calories per hour? Write your answer as two inequalities.

22. _____

23. *See question.*

22. At what time is the person burning at least 500 calories per hour? Write your answer as an inequality.

23. Open-ended Problem Make up an inequality that does not change the direction of the inequality sign when the inequality is being solved. Show the correct steps for finding the solution of your inequality.

Assessment Book, ALGEBRA 1: EXPLORATIONS AND APPLICATIONS
Copyright © McDougal Littell Inc. All rights reserved.

Test 19

QUIZ ON SECTIONS 5.1 THROUGH 5.2

DIRECTIONS: Write the answers in the spaces provided.

For Questions 1–3, solve each proportion.

1. $5:x = 10:25$

2. $\dfrac{z}{9-z} = \dfrac{5}{4}$

3. $\dfrac{4.5}{7.2} = \dfrac{c}{22.4}$

4. At Washington High School, two out of every three students usually buys a yearbook. How many yearbooks should the school plan for if there are 1230 students at the school?

Greg wants to design a garden and a fountain in a rectangular area in his backyard. He makes a drawing using a scale of 1.5 in. = 1 ft. Use this information for Questions 5 and 6.

5. What length and width should Greg use on the scale drawing to represent a rectangular fountain that is 24 in. by 30 in.?

6. The garden in the scale drawing is 18 in. by 22.5 in. What are the actual dimensions of the garden?

7. Writing Explain the meaning of the words *means* and *extremes*. Make up your own example to support your explanation.

ANSWERS

1. _____

2. _____

3. _____

4. _____

5. _____

6. _____

7. *See question.* _____

NAME _____ DATE _____ SCORE _____

Test 20

DIRECTIONS: Write the answers in the spaces provided.

ANSWERS

In the figure at the right, △XYZ ~ △MPQ.
Use the figure for Questions 1–5.

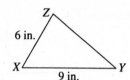

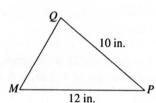

1. Find the length of $\overline{MQ}$.

2. Find the length of $\overline{YZ}$.

3. Find the ratio of the perimeter of △XYZ to the perimeter of △MPQ.

4. Find the ratio of the area of △XYZ to the area of △MPQ.

5. In the figure, assume that ∠Y = 40° and ∠M = 60°. Find the measures of ∠P and ∠Q.

6. **Writing** Explain what it means to say that two figures are similar.

1. _____

2. _____

3. _____

4. _____

5. _____

6. *See question.*

7. *See question.*

7. **Open-ended Problem** A local photo lab offers two sizes of prints, 3 in. by 5 in. (standard size) and 4 in. by 6 in. (jumbo size). The price for standard size prints is 5 for $1 while the price for jumbo size prints is 4 for $1. Are these prices proportional to the sizes of the prints? Explain why or why not.

Assessment Book, ALGEBRA 1: EXPLORATIONS AND APPLICATIONS
Copyright © McDougal Littell Inc. All rights reserved.

Test 21

QUIZ ON SECTIONS 5.5 THROUGH 5.6

DIRECTIONS: Write the answers in the spaces provided.

Find the theoretical probability of each event.

1. A die is rolled and lands showing a number greater than 2.

2. A four-sided die numbered from 1 to 4 is rolled and lands showing a number less than 4.

3. A spinner with eight equal sectors numbered from 1 to 8 is spun and stops showing a number divisible by either 2 or 3.

For Questions 4 and 5, find the theoretical probability that a randomly thrown dart that hits each target lands in the shaded area.

4.

5.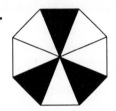

6. **Writing** Explain the difference between experimental probability and theoretical probability.

ANSWERS

1. _____

2. _____

3. _____

4. _____

5. _____

6. *See question.* _____

Test 22

TEST ON CHAPTER 5   (FORM A)

DIRECTIONS: Write the answers in the spaces provided.

Convert each ratio to a fraction in lowest terms.

1. 16 in. to 4.5 ft

2. 12 h to 3 days

Solve each proportion.

3. $\dfrac{15}{32} = \dfrac{6.6}{x}$

4. $\dfrac{3.02}{4.53} = \dfrac{n}{6}$

5. $\dfrac{2}{3} = \dfrac{x-4}{x}$

6. $\dfrac{5}{d} = \dfrac{15}{d+2}$

7. $\dfrac{3m+9}{7m+1} = \dfrac{3}{5}$

8. $\dfrac{4}{c+3} = \dfrac{28}{5(c+5)}$

Mrs. Clive wants to add an office onto her house. She draws a floor plan using a scale of 1.5 in. = 1 foot. Use this information for Questions 9 and 10.

9. What length and width should Mrs. Clive use on the floor plan to represent a rectangular room with dimensions 9 ft by 12 ft?

10. On the floor plan, the room has a closet that measures 10.5 in. by 4.5 in. What are the actual dimensions of the closet?

11. Open-ended Problem The table at the right gives the number of cats in Washington City for various years. Draw a graph to model the data. Use scales on the axes to convince readers that people need to be educated about controlling the cat population. Explain how you chose the scales for the axes.

Year	Number of cats
1980	2000
1984	2200
1988	2500
1992	2750
1996	3000

ANSWERS

1. _____

2. _____

3. _____

4. _____

5. _____

6. _____

7. _____

8. _____

9. _____

10. _____

11. *See question.* _____

Test 22

(CONTINUED)

DIRECTIONS: Write the answers in the spaces provided.

12. Writing Explain why these two figures are similar.

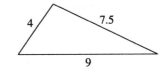

13. Find the lengths of the unknown sides in these similar triangles.

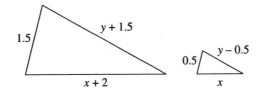

For Questions 14–16, use the fact that ABCD ~ JKLM.

14. Find *BC*.

15. The perimeter of *JKLM* is 25.5. Find the perimeter of *ABCD*.

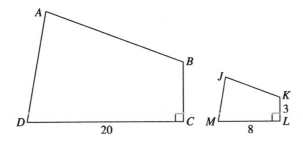

16. The area of *ABCD* is 170. Find the area of *JKLM*.

The table below shows the results of 50 spins of a spinner with sectors of equal size numbered from 1 through 8. Use this information for Questions 17 and 18.

Times spun	7	10	5	6	2	4	9	7
Number spun	1	2	3	4	5	6	7	8

17. Find the theoretical and experimental probabilities of spinning a 3 or a 4 on one spin.

18. Find the theoretical and experimental probabilities of spinning a number less than 6 on one spin.

19. Darts are randomly tossed at the square board shown at the right. What is the geometric probability that a dart which hits the board lands outside the circle?

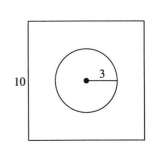

ANSWERS

12. *See question.* _____

13. *See question.* _____

14. _____

15. _____

16. _____

17. _____

18. _____

19. _____

Test 23 ••••••••••••••••••••••••••••••••••••••

TEST ON CHAPTER 5   **(FORM B)**

DIRECTIONS: Write the answers in the spaces provided.

Convert each ratio to a fraction in lowest terms.

1. 8 in. to 2.5 ft

2. 9 h to 3 days

Solve each proportion.

3. $\dfrac{9}{24} = \dfrac{7.5}{x}$

4. $\dfrac{3.69}{4.92} = \dfrac{n}{6}$

5. $\dfrac{4}{5} = \dfrac{x-3}{x}$

6. $\dfrac{3}{m} = \dfrac{9}{m+6}$

7. $\dfrac{4w+3}{9w+1} = \dfrac{1}{2}$

8. $\dfrac{7}{r+5} = \dfrac{28}{3(r+3)}$

Mr. Chuen wants to have a swimming pool and spa built in his backyard. He draws a plan using a scale of 0.5 in. = 1 foot. Use this information for Questions 9 and 10.

9. What length and width should Mr. Chuen use on the scale drawing to represent a rectangular pool with dimensions 18 ft by 35 ft?

10. On the scale drawing, the spa measures 2.5 in. by 4.25 in. What are the actual dimensions of the spa?

11. Open-ended Problem The table at the right gives the number of cars using a freeway in Washington City for various years. Draw a graph to model the data. Use scales on the axes to convince readers that people need to support alternative modes of transportation. Explain how you chose the scales for the axes.

Year	Number of cars
1992	200,000
1993	210,000
1994	230,000
1995	280,000
1996	290,000

ANSWERS

1. _____

2. _____

3. _____

4. _____

5. _____

6. _____

7. _____

8. _____

9. _____

10. _____

11. *See question.*

Test 23

(CONTINUED)

DIRECTIONS: Write the answers in the spaces provided.

ANSWERS

12. **Writing** Explain why these two figures are *not* similar.

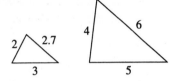

12. *See question.* _____

13. Find the lengths of the unknown sides in these similar triangles.

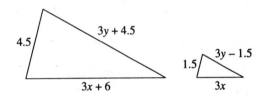

13. *See question.* _____

14. _____

15. _____

16. _____

For Questions 14–16, use the fact that EFGH ~ RSTU.

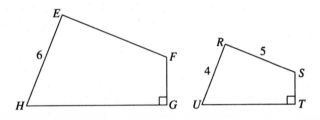

17. _____

18. _____

14. Find *EF*.

19. _____

15. The perimeter of *RSTU* is 23.4. Find the perimeter of *EFGH*.

16. The area of *EFGH* is 47.7. Find the area of *RSTU*.

The table below shows the results of 60 spins of a spinner with sectors of equal size numbered from 1 through 7. Use this information for Questions 17 and 18.

Times spun	5	12	7	8	11	7	10
Number spun	1	2	3	4	5	6	7

17. Find the theoretical and experimental probabilities of spinning a 2 or a 5 on one spin.

18. Find the theoretical and experimental probabilities of spinning a number less than 5 on one spin.

19. Darts are randomly tossed at the square board shown at the right. What is the geometric probability that a dart which hits the board lands outside the circle?

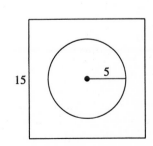

Test 24

DIRECTIONS: Write the answers in the spaces provided.

ANSWERS

The legs of a right triangle have lengths *a* and *b*. The hypotenuse has length *c*. Find the unknown length for each right triangle.

1. $a = 8, c = 17$　　**2.** $a = 5, b = 9$　　**3.** $b = 10, c = 26$

Write each decimal as a fraction in lowest terms.

4. 5.88　　**5.** $4.818181\ldots$　　**6.** $3.\overline{2}$

For Questions 7–9, simplify each expression.

7. $\sqrt{32}$　　**8.** $6 + 3\sqrt{18}$　　**9.** $\sqrt{5} \cdot \sqrt{50}$

10. Open-ended Problem Draw a right triangle. Label the lengths of the legs *c* and *m*, and the length of the hypotenuse *k*. Write the Pythagorean theorem for your triangle using the variables *c*, *k*, and *m*.

1. _____

2. _____

3. _____

4. _____

5. _____

6. _____

7. _____

8. _____

9. _____

10. *See question.* _____

Test 25

QUIZ ON SECTIONS 6.4 THROUGH 6.5

DIRECTIONS: Write the answers in the spaces provided.

For Questions 1–9, find each product.

1. $3c(c + 4)$ **2.** $6(4 + \sqrt{2})$ **3.** $(m + 3)(m - 3)$

4. $(3 - \sqrt{14})^2$ **5.** $5x(3x - 5)$ **6.** $(y - 3)^2$

7. $6p(5p)$ **8.** $(3x - 2)(x + 1)$ **9.** $(r + 2\sqrt{3})^2$

10. Write a variable expression for the area of a square with sides of length $(m + 2)$ units. Write your expression as a trinomial.

11. Write variable expressions for the perimeter and the area of the rectangle at the right.

$x + 2$

$x + 1$

12. Writing Explain the meaning of the words *monomial*, *binomial*, and *trinomial*. Give an example of each.

ANSWERS

1. _____
2. _____
3. _____
4. _____
5. _____
6. _____
7. _____
8. _____
9. _____
10. _____
11. *See question.* _____
12. *See question.* _____

Test 26

TEST ON CHAPTER 6   **(FORM A)**

DIRECTIONS: Write the answers in the spaces provided.

ANSWERS

Find the missing side length of each right triangle.

1.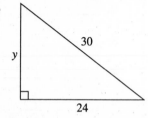

2.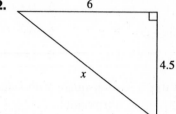

Tell whether the given lengths can be the sides of a right triangle.

3. 7.5 m, 18 m, 19.5 m   **4.** 3 ft, 6 ft, 7 ft

For Questions 5–7, write each decimal as a fraction in lowest terms.

5. 0.385   **6.** $0.\overline{93}$ **7.** $0.\overline{3}$

8. What unusual result occurs when writing the repeating decimal $2.\overline{9}$ as a fraction in lowest terms?

Write each expression in simplest form.

9. $\sqrt{96}$ **10.** $4\sqrt{3}(\sqrt{20})$ **11.** $\sqrt{75} + \sqrt{12}$

For Questions 12–14, estimate each number within a range of two integers.

12. $\sqrt{42}$ **13.** $\sqrt{119}$ **14.** $\sqrt{78}$

15. Writing Seth Roberts and his construction crew are adding a recreation room to a house. They have used a tape measure to determine if two walls are at right angles to each other. Their measurements of the sides of a triangle in the corner between the two walls are 12 in., 12 in., and 17.4 in. Do the two walls meet at a right angle? Explain your answer.

1. _____

2. _____

3. _____

4. _____

5. _____

6. _____

7. _____

8. *See question.* _____

9. _____

10. _____

11. _____

12. _____

13. _____

14. _____

15. *See question.* _____

Test 26

(CONTINUED)

DIRECTIONS: Write the answers in the spaces provided.

ANSWERS

16. Find the perimeter, the area, and the length of a diagonal of the rectangle shown at the right. Express each answer in simplest form.

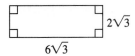
$6\sqrt{3}$, $2\sqrt{3}$

16. *See question.*

17. _____

18. _____

Find each product.

17. $6(3x + 2)$ 18. $5k(k - 3)$ 19. $(3 + 2\sqrt{7})(3 - 2\sqrt{7})$

20. $\sqrt{5}(7 + \sqrt{5})$ 21. $(2x + 3)(3x - 4)$ 22. $(2 + 3\sqrt{12})(4 + 5\sqrt{27})$

19. _____

20. _____

For Questions 23 and 24, refer to the figure at the right.

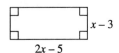
$x - 3$, $2x - 5$

23. Write a trinomial that represents the area of the rectangle.

24. What is the smallest integer value of x that will give positive values for the length and the width of the rectangle?

For Questions 25–30, find each product.

25. $(k - 7)^2$ 26. $(3x - 5)^2$ 27. $\left(7x - \dfrac{1}{2}\right)\left(7x + \dfrac{1}{2}\right)$

28. $(x + 8)(x - 8)$ 29. $(5 + 2\sqrt{3})^2$ 30. $(6 + \sqrt{6})(6 - \sqrt{6})$

31. **Writing** Explain how to use the product of a sum and a difference to find the product of $3\dfrac{3}{4}$ and $4\dfrac{1}{4}$ mentally.

21. _____

22. _____

23. _____

24. _____

25. _____

26. _____

27. _____

28. _____

29. _____

30. _____

31. *See question.*

Test 27

TEST ON CHAPTER 6   (FORM B)

DIRECTIONS: Write the answers in the spaces provided.

ANSWERS

Find the missing side length of each right triangle.

1.
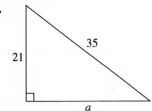

2.
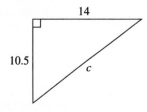

Tell whether the given lengths can be the sides of a right triangle.

3. 6 cm, 10 cm, 12 cm   **4.** 12.5 yd, 30 yd, 32.5 yd

For Questions 5–7, write each decimal as a fraction in lowest terms.

5. 0.835   **6.** $0.\overline{89}$ **7.** $0.\overline{6}$

8. What unusual result occurs when writing the repeating decimal $7.\overline{9}$ as a fraction in lowest terms?

Write each expression in simplest form.

9. $\sqrt{128}$ **10.** $3\sqrt{5}(\sqrt{20})$ **11.** $\sqrt{27} + \sqrt{75}$

For Questions 12–14, estimate each number within a range of two integers.

12. $\sqrt{62}$ **13.** $\sqrt{129}$ **14.** $\sqrt{84}$

15. Writing Beth Roberts and her construction crew are adding a recreation room to a house. They have used a tape measure to determine if two walls are at right angles to each other. Their measurements of the sides of a triangle in the corner between the two walls are 10 in., 10 in., and 13.4 in. Do the two walls meet at a right angle? Explain your answer.

1. _____

2. _____

3. _____

4. _____

5. _____

6. _____

7. _____

8. *See question.* _____

9. _____

10. _____

11. _____

12. _____

13. _____

14. _____

15. *See question.* _____

Test 27

(CONTINUED)

DIRECTIONS: Write the answers in the spaces provided.

ANSWERS

16. Find the perimeter, the area, and the length of a diagonal of the rectangle shown at the right. Express each answer in simplest form.

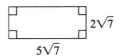
$5\sqrt{7}$ $2\sqrt{7}$

16. *See question.* _____

17. _____

18. _____

Find each product.

17. $5(7x + 3)$ **18.** $7m(m - 6)$ **19.** $(6 + 5\sqrt{3})(6 - 5\sqrt{3})$

20. $\sqrt{7}(5 + \sqrt{7})$ **21.** $(3x + 4)(4x - 5)$ **22.** $(5 + 2\sqrt{18})(3 + 3\sqrt{8})$

19. _____

20. _____

For Questions 23 and 24, refer to the figure at the right.

$2x - 7$ $x - 4$

23. Write a trinomial that represents the area of the rectangle.

24. What is the smallest integer value of x that will give positive values for the length and the width of the rectangle?

21. _____

22. _____

23. _____

24. _____

For Questions 25–30, find each product.

25. $(d - 9)^2$ **26.** $(3x - 7)^2$ **27.** $\left(8x - \frac{1}{5}\right)\left(8x + \frac{1}{5}\right)$

28. $(m + 7)(m - 7)$ **29.** $(7 + 2\sqrt{5})^2$ **30.** $(7 + \sqrt{7})(7 - \sqrt{7})$

25. _____

26. _____

31. Writing Explain how to use the product of a sum and a difference to find the product of 9.8 and 10.2 mentally.

27. _____

28. _____

29. _____

30. _____

31. *See question.* _____

NAME _____ DATE _____ SCORE _____

Test 28

DIRECTIONS: Write the answers in the spaces provided.

ANSWERS

Evaluate each expression for $x = 3$, $y = -6$, and $z = 8$.

1. $3x + 2y$ **2.** $6z^2 - 3y$ **3.** $\dfrac{x^2 - z^2}{y}$

Simplify each expression.

4. $-5 + |-7|$ **5.** $(-7) - (-1)$ **6.** $(-6)(-9)$

7. $8x - 9 + 6x + 41$ **8.** $-7(4x - y) + 6(3x - 7y)$

Mieko is selling T-shirts and caps at a school fundraiser. The T-shirts sell for \$12 each and the caps sell for \$4 each.

9. Write a variable expression that represents Mieko's income if each of her friends buys one T-shirt and one cap. Let n = the number of her friends buying a T-shirt and cap.

10. What will Mieko's income be if she sells one T-shirt and one cap to each of 15 friends?

The Greenfields are planning their daughter's graduation party. They have agreed to pay \$450 to rent a banquet hall and the caterer they have chosen charges \$22 per person. Use this information for Questions 11 and 12.

11. Write an equation describing the Greenfields' cost for the banquet hall and catering as a function of the number of people at the party.

12. The Greenfields have budgeted \$2000 for the banquet hall and catering expense. What is the maximum number of people who can attend the party for this amount?

13. Which equation below is an identity? Which has no solution?
A. $2(6x + 4) + 8 = 12x + 30$ B. $12x - 24 = 5(16x + 10)$
C. $10(c + 3) - 30 = 12c + 36$ D. $4(6a + 3) = 3(8a + 4)$

14. What is the smallest number that you can multiply by to eliminate the fractions in the equation $\frac{1}{2} + \frac{9}{10}m = \frac{4}{15} - \frac{1}{6}m$?

Solve each proportion.

15. $\dfrac{7}{d} = \dfrac{28}{d + 2}$ **16.** $\dfrac{4}{5} = \dfrac{9}{x}$ **17.** $\dfrac{2.5}{3.5} = \dfrac{x - 4}{x}$

1. _____

2. _____

3. _____

4. _____

5. _____

6. _____

7. _____

8. _____

9. _____

10. _____

11. _____

12. _____

13. _____

14. _____

15. _____

16. _____

17. _____

Test 28

(CONTINUED)

DIRECTIONS: Write the answers in the spaces provided.

Solve each equation. If an equation is an *identity* or if there is *no solution*, say so.

18. $-4r + 2.6 = 10.6$ **19.** $\frac{x}{2} - 8 = 2$ **20.** $-p = 43$

21. $\frac{4x}{7} - 12 = -16$ **22.** $7k = 9k + 5$ **23.** $8 - 6m = 4m - 7$

24. $\frac{3}{5}(10x - 15) = 6x + 5$ **25.** $\frac{3}{4}v - \frac{1}{6} = \frac{2}{3}v + \frac{7}{8}$

26. $-5w - 8 = 8(2w - 3)$ **27.** $0.8 - 0.07h = 0.92 + 0.53h$

Solve each inequality.

28. $7y + 9 > 51$ **29.** $3(x - 4) < 2x + 3$

For Questions 30 and 31, find the slope of the line passing through each pair of points.

30. $(10, 5)$ and $(-3, 7)$ **31.** $(-6, 1)$ and $(3, -2)$

32. Writing Explain the difference between positive slope and negative slope, and between zero slope and undefined slope. Include graphs showing lines with each of these slopes.

ANSWERS

18. _____

19. _____

20. _____

21. _____

22. _____

23. _____

24. _____

25. _____

26. _____

27. _____

28. _____

29. _____

30. _____

31. _____

32. *See question.* _____

Test 28

(CONTINUED)

DIRECTIONS: Write the answers in the spaces provided.

33. Find an equation of a line with slope $\frac{1}{2}$ passing through $(-4, 2)$.

Find an equation of the line through each pair of points.

34. $(-4, -1)$ and $(2, -2)$

35. $(5, 4)$ and $(-2, 4)$

Graph each equation.

36. $y = -2x$

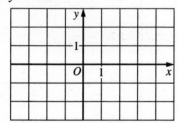

37. $y = 3x - 1$

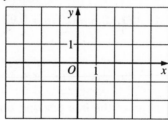

For Questions 38 and 39, express each rate in the given units.

38. $\dfrac{\$15.60}{4 \text{ ft}} = \underline{\ ?\ }$ cents/in.

39. $\dfrac{3.5 \text{ mi}}{50 \text{ min}} = \underline{\ ?\ }$ mi/h

40. Sanaia is considering buying a season ticket to the opera. A season ticket for an orchestra seat costs $550 for 10 operas. The cost of buying a ticket for the same seat for one opera is $72. Write and solve an equation to determine how many operas Sanaia will have to attend in order to recover the cost of a season ticket.

41. Writing Explain the meaning of geometric probability and how it relates to the target shown at the right and to the question below.

What is the geometric probability that a dart which hits the target will land in the shaded square?

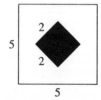

ANSWERS

33. _____

34. _____

35. _____

36. *See question.* _____

37. *See question.* _____

38. _____

39. _____

40. *See question.* _____

41. *See question.* _____

Assessment Book, ALGEBRA 1: EXPLORATIONS AND APPLICATIONS

Test 28

(CONTINUED)

DIRECTIONS: Write the answers in the spaces provided.

Bill wants to build an electric train layout for his train collection. He makes a drawing using a scale of 2 in. = 1 ft.

42. What length and width should Bill use on the scale drawing to represent a rectangular layout with dimensions 4 ft by 7 ft?

43. On the scale drawing, the layout has a train station that measures 2.5 in. by 3 in. What are the actual dimensions of the train station?

The ratio of the lengths of corresponding sides of two similar pentagons is 7:2.

44. If the perimeter of the larger pentagon is 84 cm, what is the perimeter of the smaller pentagon?

45. If the area of the smaller pentagon is 12 cm^2, what is the area of the larger pentagon?

Tell whether the given lengths can be the sides of a right triangle.

46. 8.4 m, 6.3 m, 10.5 m   **47.** 5 ft, 7 ft, 8 ft

Write each expression in simplest form.

48. $\sqrt{76}$ **49.** $4\sqrt{10}(\sqrt{40})$ **50.** $\sqrt{98} + \sqrt{50}$

For Questions 51–55, find each product.

51. $(5x + 3)(3x - 2)$ **52.** $(x + 7)(x - 7)$ **53.** $(2x - 5)^2$

54. $(5 + \sqrt{5})(5 - \sqrt{5})$ **55.** $(2 + 3\sqrt{12})(4 + 5\sqrt{27})$

56. Open-ended Problem Using the variable x, write binomial expressions for the length and the width of a rectangle. Draw and label your rectangle. Write a trinomial expression for the area of the rectangle. What is the smallest integer value of x that will give positive values for the length and the width of the rectangle?

ANSWERS

42. _____

43. _____

44. _____

45. _____

46. _____

47. _____

48. _____

49. _____

50. _____

51. _____

52. _____

53. _____

54. _____

55. _____

56. *See question.* _____

Test 29

DIRECTIONS: Write the answers in the spaces provided.

ANSWERS

Give the *x*- and *y*-intercepts of each graph. Then graph each equation.

1. $3x + 5y = -15$

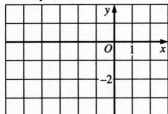

2. $-2x + 7y = 14$

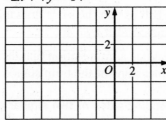

Rewrite each equation in slope-intercept form. Then graph each equation.

3. $3x - y = 5$

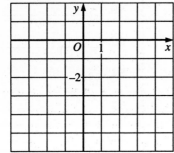

4. $4x + 3y = 12$

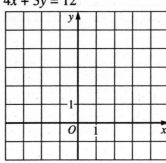

Use substitution to solve each system of equations.

5. $4x + 3y = 11$
$y = x - 1$

6. $5k - 4m = 20$
$k - 2m = 16$

On a history test, Reggie receives three points for each correct answer and loses one point for each incorrect answer.

7. Write an equation showing that Reggie received 60 points.

8. Write an equation showing that Reggie answered all 40 questions on the test.

9. Solve the system of equations you wrote for Questions 7 and 8 by graphing.

10. Writing Explain the meaning of the point of intersection of the two lines graphed in Question 9 in terms of Reggie's score.

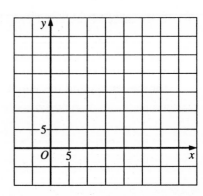

1. _____

2. _____

3. _____

4. _____

5. _____

6. _____

7. _____

8. _____

9. _____

10. *See question.* _____

Test 30

QUIZ ON SECTIONS 7.3 THROUGH 7.4

DIRECTIONS: Write the answers in the spaces provided.

ANSWERS

For Questions 1–6, solve each system of equations.

1. $3x + 2y = 18$
$4x + 2y = 4$

2. $5x + 7y = 12$
$-5x - 3y = 8$

3. $x - y = -4$
$-5x + 2y = 2$

4. $2x - 3y = 24$
$3x - 2y = 21$

5. $2x - 5y = 7$
$3x - 2y = -17$

6. $3y + 9x = 18$
$-4 + 2y = 6x$

7. Shareh and Ranil work for an appliance store that pays an hourly wage plus a commission for each new appliance they sell. One week Shareh worked for 40 h and sold 20 appliances for a total pay of $400. Ranil worked 36 h during the same week, but only sold 7 appliances for a total pay of $250. What is the hourly wage paid and what is the commission paid on each appliance sold?

8. Open–ended Problem Which method of solving a system of equations do you prefer to use for the system at the right? Explain why you prefer this method, and then use the method to solve the system.

$4x - y = 12$
$2x - y = 6$

1. _____

2. _____

3. _____

4. _____

5. _____

6. _____

7. *See question.*
_____

8. *See question.*
_____

NAME _____ DATE _____ SCORE _____

Test 31

DIRECTIONS: Write the answers in the spaces provided.

ANSWERS

Graph each inequality or system of inequalities.

1. $2x - 3y \leq -6$

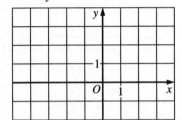

2. $4x - 3y > 6$

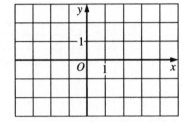

3. $2x + y \leq 1$
$2x - y \geq -3$

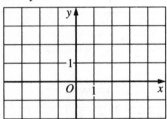

4. $2x - y > -2$
$3x + 2y \leq 6$

1. See question. _____

2. See question. _____

3. See question. _____

4. See question. _____

5. See question. _____

6. See question. _____

7. _____

8. See question. _____

The Pioneer Math Club must sell at least 10 school jackets and at least 20 caps during a fundraiser to buy calculators. The club will make $15 profit on every jacket sold and $2 profit on every cap sold. Use this information for Questions 5–7.

5. Write a system of inequalities that shows how many jackets and caps the club members need to sell to make a profit of at least $300.

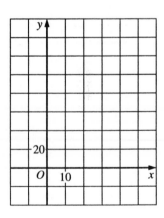

6. Graph the system of inequalities you wrote for Question 5.

7. Will the club meet its goal of $300 profit if it sells 14 jackets and 37 caps? How much above or below its goal will the club be?

8. Open–ended Problem Write a system of inequalities that will form a triangular region when graphed. Then sketch the graph of your system.

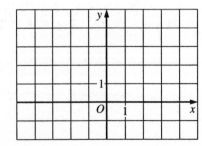

Assessment Book, ALGEBRA 1: EXPLORATIONS AND APPLICATIONS

Test 32

TEST ON CHAPTER 7   **(FORM A)**

DIRECTIONS: Write the answers in the spaces provided.

Give the horizontal intercept and the vertical intercept of each equation. Then graph each equation.

1. $2x - 3y = 6$

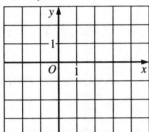

2. $4x + 5y = 20$

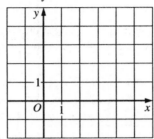

Susanna is packaging a blend of mixed nuts and candies for wedding favors. The mixed nuts cost her $3 per pound and the candies cost her $4 per pound. Use this information for Questions 3–5.

3. Write an equation showing that Susanna spends $27 altogether for the nuts and candies.

4. Write an equation showing that Susanna bought a total of 8 lb of nuts and candies.

5. Solve the system of equations to find how many pounds of nuts and how many pounds of candies Susanna bought.

For Questions 6–11, solve each system of equations.

6. $4x + 3y = 16$
$2x + y = 6$

7. $5x - 2y = -15$
$x - 2y = -3$

8. $2x - 7y = 3$
$3x - 7y = 8$

9. $3x + 7y = -4$
$-2x - 4y = 4$

10. $4x - y = -7$
$3x - 2y = 1$

11. $2x + 5y = 3$
$x - 2y = -3$

12. Writing Without graphing, tell which pair(s) of equations below make a system of equations with one solution. Explain how you know.

A. $5x - y = 3$ B. $3y - 6x = -12$ C. $3y + 2 = 6x$

ANSWERS

1. _____

2. _____

3. _____

4. _____

5. *See question.*

6. _____

7. _____

8. _____

9. _____

10. _____

11. _____

12. *See question.*

NAME _____ DATE _____ SCORE _____

Test 32

DIRECTIONS: Write the answers in the spaces provided.

Graph each inequality or system of inequalities.

13. $y < x + 3$

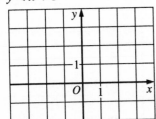

14. $4y > 3x - 4$

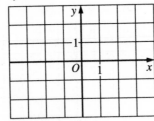

15. $y \leq 2$
$y \geq -3x$

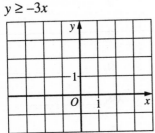

16. $2x + 3y > 6$
$3x - 4y \geq 4$

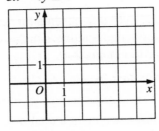

13. *See question.* _____

14. *See question.* _____

15. *See question.* _____

16. *See question.* _____

17. _____

18. _____

19. *See question.* _____

20. _____

21. _____

For a fundraiser, Pioneer High School can purchase at most 300 coupon books which are offered in two prices. The sports events book costs $15 and the restaurant book costs $25. Pioneer High School can spend no more than $6000 toward the purchase of the coupon books. Use this information for Questions 17–20.

17. Write an inequality showing that the sum of x restaurant books and y sports events books comes to a total of at most 300 books.

18. Write an inequality showing that Pioneer High School can spend up to $6000 on the coupon books.

19. Since they cannot buy a negative number of coupon books, $x \geq 0$ and $y \geq 0$. Graph these inequalities and the inequalities you wrote for Questions 17 and 18 in the same coordinate plane.

20. What is the greatest number of restaurant coupon books Pioneer High School can buy?

21. At a high school football game, 500 tickets were sold. Adult tickets cost $5 and student tickets cost $3. If the total amount collected was $2300, how many student tickets were sold?

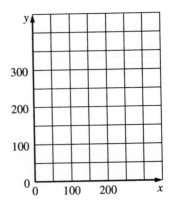

Assessment Book, ALGEBRA 1: EXPLORATIONS AND APPLICATIONS
Copyright © McDougal Littell Inc. All rights reserved.

48

Test 33

TEST ON CHAPTER 7   **(FORM B)**

DIRECTIONS: Write the answers in the spaces provided.

ANSWERS

Give the horizontal intercept and the vertical intercept of each equation. Then graph each equation.

1. $3x - 2y = 6$

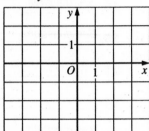

2. $5x + 4y = 20$

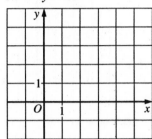

1. _____

2. _____

3. _____

4. _____

5. *See question.* _____

6. _____

7. _____

8. _____

9. _____

10. _____

11. _____

12. *See question.* _____

Tyrone is packaging a mix of bluegrass seed and drought-resistant seed for people buying grass seed for their lawns. The bluegrass seed costs him $2 per pound while the drought-resistant grass seed costs him $3 per pound. Use this information for Questions 3–5.

3. Write an equation showing that Tyrone spent $68 altogether for the two types of grass seed.

4. Write an equation showing that Tyrone bought a total of 25 lb of the two types of grass seed.

5. Solve the system of equations to find out how many pounds of each type of grass seed Tyrone bought.

For Questions 6–11, solve each system of equations.

6. $3x + 2y = 12$
   $x - y = -1$

7. $2x + 3y = -1$
   $2x - 7y = -11$

8. $x + 3y = 5$
   $6x + 3y = 0$

9. $x + y = 6$
   $2x - 3y = -13$

10. $3x + 4y = 12$
   $6x - 8y = -24$

11. $3x - 4y = -4$
   $-2x + 5y = -2$

12. Writing Without graphing, tell which pair(s) of equations below make a system of equations with no solutions. Explain how you know.

   A. $4y - 8x = -6$ B. $3x - 4y = 2$ C. $4y + 2 = 8x$

Test 33

(CONTINUED)

DIRECTIONS: Write the answers in the spaces provided.

Graph each inequality or system of inequalities.

13. $y < x + 2$

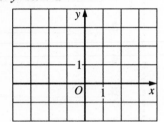

14. $3x - 2y < 6$

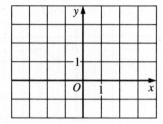

15. $y \le 3$
$y \ge -2x$

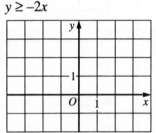

16. $3x + 4y < 8$
$2x - 3y \ge 3$

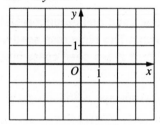

ANSWERS

13. *See question.*

14. *See question.*

15. *See question.*

16. *See question.*

17. _____

18. _____

19. *See question.*

20. _____

21. _____

For a party, Maria and Bettina need to buy no more than 100 drinks. At a discount store, sodas cost 30 cents each and sparkling water drinks cost 40 cents each. Maria and Bettina can spend up to $35 on the drinks. Use this information for Questions 17–20.

17. Write an inequality showing that the sum of x cans of soda and y bottles of sparkling water is at most 100 drinks.

18. Write an inequality showing that Maria and Bettina can spend up to $35 on the drinks.

19. Since they cannot buy a negative numbers of drinks, $x \ge 0$ and $y \ge 0$. Graph these inequalities and the inequalities you wrote for Questions 17 and 18 in the same coordinate plane.

20. What is the greatest number of sparking water drinks Maria and Bettina can buy?

21. At a high school basketball game, 400 tickets were sold. Adult tickets cost $5 and student tickets cost $2.50. If the total amount collected was $1375, how many student tickets were sold?

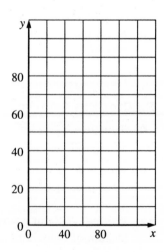

NAME _____ DATE _____ SCORE _____

Test 34

QUIZ ON SECTIONS 8.1 THROUGH 8.3

DIRECTIONS: Write the answers in the spaces provided.

Graph each equation. Tell whether each equation is *linear* or *nonlinear*.

1. $y = 3|x| - 2$

2. $y = -2x^2 + 3$

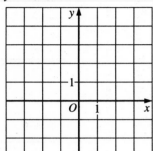
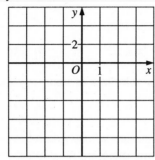

Predict how the graph of each equation will compare with the graph of $y = x^2$.

3. $y = 3x^2$

4. $y = -0.5x^2$

ANSWERS

1. _____

2. _____

3. *See question.* _____

4. *See question.* _____

5. _____

6. _____

7. _____

8. _____

9. _____

10. _____

11. *See question.* _____

For Questions 5–10, solve each equation algebraically.

5. $x^2 = 121$

6. $\frac{1}{2}m^2 = 4$

7. $0 = k^2 - 9$

8. $-9c^2 + 36 = 0$

9. $3r^2 + 6 = 66$

10. $5x^2 + 64 = 74$

11. Open-ended Problem Make up an equation of (**a**) a line that will cross the *x*-axis at $x = 1$, (**b**) a parabola that will touch the *x*-axis exactly once at $x = 1$, and (**c**) a parabola that will cross the *x*-axis at both $x = 1$ and $x = -1$. Graph all your equations on the same set of axes.

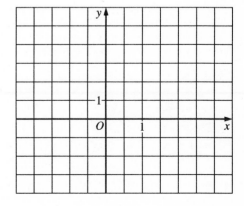

Test 35

DIRECTIONS: Write the answers in the spaces provided.

ANSWERS

Solve each equation using a graphing calculator.

1. $x^2 = 7x - 6$　　　**2.** $x^2 + 4 = 5x$　　　**3.** $x^2 + 5x = -3$

Solve each equation using the quadratic formula.

4. $x^2 + 12 = 7x$　　　**5.** $3x^2 = 7x + 6$　　　**6.** $2x^2 = x + 5$

For Questions 7–9, find the number of solutions of each equation.

7. $4x^2 = 12x - 9$　　　**8.** $4x = 5x^2 + 3$　　　**9.** $4x^2 + 1 = 5x$

10. Open-ended Problem Write an equation whose graph has no
x-intercepts. Explain how you know there are no x-intercepts and then
graph your equation.

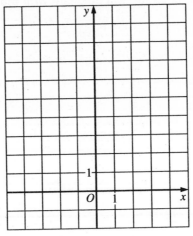

1. _____

2. _____

3. _____

4. _____

5. _____

6. _____

7. _____

8. _____

9. _____

10. _See question._

Test 36

TEST ON CHAPTER 8   (FORM A)

DIRECTIONS: Write the answers in the spaces provided.

ANSWERS

Tell whether each relationship is *linear* or *nonlinear*.

1. $y = 3x + 4$ **2.** $y = 2x^2 - 1$ **3.** $y = 3|x| - 2$

4. the relationship between the radius and the area of a circle

5. the relationship between the length of a side and the perimeter of a square

Match each equation with its graph.

6. $y = 2x^2$

7. $y = -3x^2$

8. $y = \frac{1}{5}x^2$

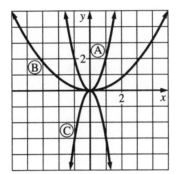

Describe how the graph of each equation compares to the graph of $y = x^2$.

9. $y = 3x^2$

10. $y = -4x^2$

11. $y = -\frac{1}{6}x^2$

1. _____

2. _____

3. _____

4. _____

5. _____

6. _____

7. _____

8. _____

9. *See question.*

10. *See question.*

11. *See question.*

12. _____

13. _____

14. _____

15. _____

16. _____

17. _____

Solve each equation algebraically.

12. $x^2 + 9 = 18$ **13.** $2x^2 = 72$ **14.** $4x^2 - 4 = 60$

15. $\frac{1}{3}x^2 + 4 = 7$ **16.** $\frac{3}{4}x^2 - 9 = 39$ **17.** $3x^2 - 42 = 201$

Assessment Book, ALGEBRA 1: EXPLORATIONS AND APPLICATIONS

53

Test 36

(CONTINUED)

DIRECTIONS: Write the answers in the spaces provided.

ANSWERS

18. Estimate the solutions of the equation $3x^2 - 8 = 0$ using the related graph shown at the right.

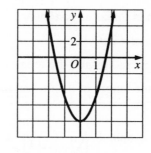

18. _____

19. _____

20. _____

21. _____

Solve each equation using a graphing calculator.

19. $3x^2 - 5x = 1$ **20.** $2x^2 + 1 = -9x$ **21.** $5x^2 + 4x + 6 = 0$

22. _____

Solve each equation using the quadratic formula.

23. _____

22. $x^2 + 6x = -5$ **23.** $81x^2 = 18x - 1$ **24.** $6x - 1 = 3x^2$

24. _____

Find the number of solutions of each equation.

25. $3x^2 + 9x = 2$ **26.** $5x^2 - x + 7 = 0$ **27.** $3x^2 + 6x = -3$

25. _____

26. _____

For Questions 28–30, use the figures at the right. The two figures have the same area.

28. Write an equation showing that the square and the rectangle have the same area.

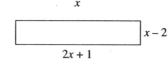

27. _____

29. How many solutions will your equation from Question 28 have?

28. _____

30. Solve your equation. Do all the solutions make sense in this situation?

29. _____

30. *See question.*

31. Writing Explain what the discriminant is and why it is useful.

31. *See question.*

NAME _____ DATE _____ SCORE _____

Test 37

TEST ON CHAPTER 8   **(FORM B)**

DIRECTIONS: Write the answers in the spaces provided.

ANSWERS

Tell whether each relationahip is _linear_ or _nonlinear._

1. $y = 3x^2 + 5$ **2.** $y = 4|x| + 3$ **3.** $y = 4x + 3$

4. the relationship between the diameter and the circumference of a circle

5. the relationship between the radius and the area of a circle

Match each equation with its graph.

6. $y = 3x^2$

7. $y = \frac{1}{3}x^2$

8. $y = -4x^2$

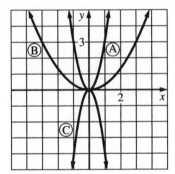

Describe how the graph of each equation compares to the graph of $y = x^2$.

9. $y = -\frac{5}{2}x^2$

10. $y = 4x^2$

11. $y = 0.5x^2$

1. _____

2. _____

3. _____

4. _____

5. _____

6. _____

7. _____

8. _____

9. _See question._

10. _See question._

11. _See question._

12. _____

13. _____

14. _____

15. _____

16. _____

17. _____

Solve each equation algebraically.

12. $x^2 - 6 = 30$ **13.** $3x^2 = 27$ **14.** $5x^2 - 15 = 110$

15. $\frac{1}{4}x^2 + 5 = 9$ **16.** $\frac{4}{25}x^2 - 9 = 55$ **17.** $4x^2 - 95 = 101$

Test 37

(CONTINUED)

DIRECTIONS: Write the answers in the spaces provided.

ANSWERS

18. Estimate the solutions of the equation $4x^2 - 12 = 0$ using the related graph shown at the right.

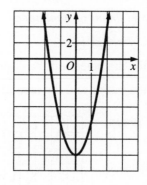

18. _____

19. _____

20. _____

21. _____

22. _____

Solve each equation using a graphing calculator.

19. $2x^2 - 4x = 3$ 20. $4x^2 + 9x = -1$ 21. $6x^2 + 4x + 7 = 0$

23. _____

Solve each equation using the quadratic formula.

22. $x^2 + 9x + 8 = 0$ 23. $64x^2 + 1 = 16x$ 24. $5x + 2 = 4x^2$

24. _____

Find the number of solutions of each equation.

25. $6x^2 + 4 = -9x$ 26. $5x^2 - 10x + 5 = 0$ 27. $4x^2 + 8x + 2 = 0$

25. _____

26. _____

For Questions 28–30, use the figures at the right. The two figures have the same area.

28. Write an equation showing that the square and the rectangle have the same area.

27. _____

28. _____

29. How many solutions will your equation from Question 28 have?

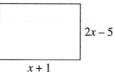

29. _____

30. Solve your equation. Do all the solutions make sense in this situation?

30. *See question.* _____

31. *See question.* _____

31. **Open-ended Problem** Make up a quadratic equation and solve it using a graphing calculator and also by using the quadratic formula. Tell which method you like better and why.

NAME _____ DATE _____ SCORE _____

Test 38

DIRECTIONS: Write the answers in the spaces provided.

ANSWERS

Evaluate each power.

1. 7^3 **2.** $(-3)^4$ **3.** $\left(\frac{1}{4}\right)^3$

Evaluate each expression when $x = 3$ and $y = 5$.

4. x^2y **5.** $(3x - y)^2$ **6.** $3x^3 - 2y^2$

Aiyisha is saving for her college tuition. She has \$2,000 in a savings account that pays 6% annual interest.

7. Write an equation giving the amount y of money in Aiyisha's account x years from now, assuming she makes no deposits or withdrawals.

8. How much money will Aiyisha have in her account 4 years from now?

During a drought, the water usage of a city decreased by an average of 4% per year since 1985 when it was 300,000 cubic hundred feet. Use this information for Questions 9 and 10.

9. Write an exponential function to model this situation.

10. Estimate the amount of water used by the city in 1992.

11. Writing Explain what the terms *growth rate* and *growth factor* mean and how they relate to exponential growth. Include an example to illustrate your explanation.

1. _____

2. _____

3. _____

4. _____

5. _____

6. _____

7. _____

8. _____

9. _____

10. _____

11. *See question.* _____

Test 39

DIRECTIONS: **Write the answers in the spaces provided.**

ANSWERS

Evaluate each power.

1. 5^{-3} **2.** 6.701^0 **3.** $\left(\frac{5}{2}\right)^{-2}$

Evaluate each expression when $x = -3$, $y = 5$, **and** $z = 4$.

4. $y^{-3}z$ **5.** x^2z^{-3} **6.** $\left(\frac{xy}{z}\right)^{-2}$

The cat population in a certain city x **years after 1990 is modeled by the function** $y = 1250(1.025)^x$. **Estimate the cat population in this city for each of these years.**

7. 1985   **8.** 1995   **9.** 2010

Write each number in decimal notation.

10. 9.07×10^5 **11.** 5.28×10^{-3} **12.** 4×10^{-5}

For Questions 13–15, write each number in scientific notation.

13. 390,000,000   **14.** 0.00000012   **15.** 82.43

16. Writing Explain what scientific notation is and why it is useful. Give examples to illustrate your explanation.

1. _____

2. _____

3. _____

4. _____

5. _____

6. _____

7. _____

8. _____

9. _____

10. _____

11. _____

12. _____

13. _____

14. _____

15. _____

16. *See question.*

NAME _____ DATE _____ SCORE _____

Test 40

DIRECTIONS: Write the answers in the spaces provided.

For Questions 1–6, simplify each expression.

1. $w^3 \cdot w^{12}$　　　　**2.** $\dfrac{n^{13}}{n^8}$　　　　**3.** $\dfrac{x^9 z^{-3}}{x^6 z^2}$

4. $(c^2 d)^5$　　　　**5.** $(r^7)^6$　　　　**6.** $\left(\dfrac{a^3 b^5}{b^2 c^4}\right)^2$

7. During a year when there were about 2.2×10^8 people in the United States, the average person consumed approximately 25 quarts of ice cream. Estimate the total number of quarts of ice cream consumed by people in the United States during this year. Write your answer in scientific notation.

8. The equation $d = 16t^2$ gives the distance d, in feet, traveled by a falling object in t seconds. Suppose rock A has been falling three times as long as rock B. Complete the following equation. Justify your answer algebraically.

(distance traveled by rock A) = ___?___ · (distance traveled by rock B)

ANSWERS

1. _____

2. _____

3. _____

4. _____

5. _____

6. _____

7. _____

8. *See question.*
_____

9. *See question.*
_____

9. Writing Explain the difference between the *product of powers* rule of exponents and the *power of a product* rule of exponents. Give an example of each to illustrate the difference between the rules and check your answer by another method.

NAME _____ DATE _____ SCORE _____

Test 41

TEST ON CHAPTER 9   **(FORM A)**

DIRECTIONS: Write the answers in the spaces provided.

Evaluate each power.

1. 3^4 **2.** $(-5)^4$ **3.** $\left(-\frac{1}{4}\right)^3$ **4.** $\left(\frac{2}{7}\right)^{-2}$

5. 0.2^3 **6.** 7.13^0 **7.** $(-2)^{-4}$ **8.** 3^{-4}

Evaluate each expression when $x = -3$ and $y = 2$.

9. $xy^2 + (3 - x)^2$ **10.** $(x + y)^3 + x^3 y$ **11.** $2x^3 - 3y^2 + (xy)^2$

Evaluate each expression when $x = 2$, $y = -3$, and $z = 4$.

12. $x^{-1}y^2$ **13.** $x^{-2}z + y^2$ **14.** $\left(\frac{xy}{z}\right)^{-1}$

$15,000 is placed in a mutual fund account that pays 11.5% interest annually. No additional funds are placed into the account and no fees are charged. Use this information for Questions 15 and 16.

15. Write an equation that gives the amount of money y in the account x years from now.

16. How much money will be in the account 7 years from now?

17. Open-ended Problem Using the rules of exponents, create an example that could be used to show that $x^0 = 1$ for $x \neq 1$. Explain your reasoning.

ANSWERS

1. _____

2. _____

3. _____

4. _____

5. _____

6. _____

7. _____

8. _____

9. _____

10. _____

11. _____

12. _____

13. _____

14. _____

15. _____

16. _____

17. *See question.* _____

Test 41

(CONTINUED)

DIRECTIONS: Write the answers in the spaces provided.

ANSWERS

A new house built in 1995 was valued at $120,000. Each year the value of the house increases by 2%.

18. Write an equation giving the value y of the house x years after 1995.

19. How much will the house be worth in the year 2005?

The mass y in grams of a colony of bacteria x hours after 9 A.M., May 19, is given by the equation $y = 1.8(1.56^x)$. Estimate the mass at each time.

20. 12 noon, May 19   **21.** 5 A.M., May 19

Write each number in decimal notation.

22. 3.24×10^5 **23.** 4.6×10^{-4} **24.** 3.045×10^4

Write each number in scientific notation.

25. 24,100,000   **26.** 0.0000078   **27.** 8.003

For Questions 28–33, simplify each expression.

28. $\dfrac{k^5 r^3}{k^3 r^6}$ **29.** $(3x^2)^5$ **30.** $(2a^4 b^5)(-3a^3 b^9)$

31. $\dfrac{x^{-5} w^3}{w^2 z^{-2}}$ **32.** $\left(\dfrac{-5a^{-2} b^4}{2a^0 b^3}\right)^3$ **33.** $\left(\dfrac{m^4 n^5}{nm^7}\right)^3$

34. Writing During a recent flight, the space shuttle traveled 2.5×10^6 mi. The total cost of the flight was $\$5 \times 10^7$. Explain how to find the cost per mile traveled using the numbers in their given form. Include the process for finding the answer and the answer in your explanation.

18. _____

19. _____

20. _____

21. _____

22. _____

23. _____

24. _____

25. _____

26. _____

27. _____

28. _____

29. _____

30. _____

31. _____

32. _____

33. _____

34. *See question.*
_____

Test 42

DIRECTIONS: Write the answers in the spaces provided.

ANSWERS

Evaluate each power.

1. 4^3 **2.** $(-3)^5$ **3.** $\left(-\frac{1}{5}\right)^4$ **4.** $\left(\frac{3}{4}\right)^{-3}$

5. 4^{-2} **6.** 9.27^0 **7.** $(-3)^{-2}$ **8.** 0.2^5

Evaluate each expression when $x = -4$ and $y = 3$.

9. $xy^3 + (2 - x)^2$ **10.** $(x + y)^3 - x^2y$ **11.** $2x^3 - 2y^2 + (xy)^2$

Evaluate each expression when $x = 3$, $y = -4$, and $z = 2$.

12. $z^{-1}y^2$ **13.** $z^{-2}x + y^2$ **14.** $\left(\frac{xz}{2y}\right)^{-1}$

$20,000 is placed in a mutual fund account that pays 9.5% interest annually. No additional funds are placed into the account and there are no fees charged. Use this information for Questions 15 and 16.

15. Write an equation giving the amount of money y in the account x years from now.

16. How much money will be in the account 5 years from now?

17. Open-ended Problem Using the rules of exponents, create an example that could be used to show that $x^0 = 1$ for $x \neq 1$. Explain your reasoning.

1. _____

2. _____

3. _____

4. _____

5. _____

6. _____

7. _____

8. _____

9. _____

10. _____

11. _____

12. _____

13. _____

14. _____

15. _____

16. _____

17. *See question.* _____

Test 42

(CONTINUED)

DIRECTIONS: **Write the answers in the spaces provided.**

ANSWERS

A new house built in 1995 was valued at $150,000. Each year the value of the house increases by 3%.

18. Write an equation that gives the value y of the house x years after 1995.

19. How much will the house be worth in the year 2004?

The mass y in grams of a colony of bacteria x hours after 7 P.M., May 22, is given by the equation $y = 1.9(1.72^x)$. Estimate the mass at each time.

20. 11 P.M., May 22   **21.** 2 P.M., May 22

Write each number in decimal notation.

22. 8.71×10^4 **23.** 9.2×10^{-5} **24.** 4.0213×10^5

Write each number in scientific notation.

25. 781,000,000   **26.** 0.0000037   **27.** 4.002

For Questions 28–33, simplify each expression.

28. $\dfrac{m^7 n^2}{m^4 n^6}$ **29.** $(4y^3)^3$ **30.** $(5c^3 d^5)(-3c^4 d^6)$

31. $\dfrac{x^{-6} w^4}{w^2 z^{-1}}$ **32.** $\left(\dfrac{-3b^{-2} a^5}{2a^0 b^2}\right)^3$ **33.** $\left(\dfrac{w^5 z^3}{zw^7}\right)^3$

34. Writing During a recent flight, the space shuttle traveled 3.5×10^6 mi. The total cost of the flight was $\$7 \times 10^7$. Explain how to find the cost per mile traveled using the numbers in their given form. Include the process for finding the answer and the answer in your explanation.

18. _____

19. _____

20. _____

21. _____

22. _____

23. _____

24. _____

25. _____

26. _____

27. _____

28. _____

29. _____

30. _____

31. _____

32. _____

33. _____

34. *See question.*

Test 43

QUIZ ON SECTIONS 10.1 THROUGH 10.2

DIRECTIONS: Write the answers in the spaces provided.

Add or subtract as indicated. Write each answer in standard form and give its degree.

1. $(5x^2 + 3x - 6) + (7x^2 - 2x + 4)$

2. $(3m^2 - 8m - 5) + (6m^2 - 3m + 8)$

3. $(7k^2 + 2k - 3) - (5k^2 - 3k - 9)$

4. $(2w^3 - 8w^2 + 6) - (7w^3 + 4w^2 - 1)$

For Questions 5–8, multiply.

5. $(x + 2)(3x + 1)$

6. $(3n - 4)(2n - 5)$

7. $8x(x^2 + 4x - 3)$

8. $(f + 4)(2f^2 - f + 7)$

9. Which of the following expressions is *not* a polynomial?

    A. $5x + 7x^3$ B. $18x^2 + \dfrac{x}{9}$ C. $\dfrac{x - 4}{2}$ D. $\dfrac{x + 3}{x}$

10. **Writing** Explain the difference between a linear polynomial and a nonlinear polynomial. Tell how to classify a polynomial as linear or nonlinear, and give an example of both types.

ANSWERS

1. _____

2. _____

3. _____

4. _____

5. _____

6. _____

7. _____

8. _____

9. _____

10. *See question.* _____

NAME _____ DATE _____ SCORE _____

Test 44

DIRECTIONS: Write the answers in the spaces provided.

ANSWERS

Solve each equation.

1. $x^2 - 6x = 0$ **2.** $5y^2 = -10y$ **3.** $(3x - 4)(x + 5) = 0$

1. _____

The equation $h = -16t^2 + 56t$ models the height h, in feet, of a golf ball t seconds after it was hit. Use this equation for Questions 4 and 5.

2. _____

4. Find the values of t for which $h = 0$.

3. _____

5. When is the ball 40 ft above the ground?

4. _____

Factor if possible. If not possible, write *not factorable.*

6. $x^2 + 5x + 6$ **7.** $x^2 - 5x - 36$ **8.** $2x^2 - 7x + 3$

5. _____

9. $9x^2 - 5x - 1$ **10.** $25x^2 - 4$ **11.** $12x^2 - x - 6$

6. _____

For Questions 12–14, solve each equation.

7. _____

12. $x^2 + 3x = 10$ **13.** $4x^2 + 16x = -15$ **14.** $6x^2 + 5x = 21$

8. _____

15. Writing Explain what the zero-product property is and why it is useful. Give an example to illustrate your explanation.

9. _____

10. _____

11. _____

12. _____

13. _____

14. _____

15. *See question.* _____

Test 45

DIRECTIONS: Write the answers in the spaces provided.

Tell whether each expression is a polynomial. Write _Yes_ or _No._ If not, explain why not.

1. $5x^4 + 8x - 1$

2. $\dfrac{3}{x^3} + 2x$

3. $17x^{-2}$

Add. Give the degree of each sum.

4. $5w^2 + 3w - 1$
   $\underline{2w^2 - \ w + 1}$

5. $-8k^3 + 2k - 5$
   $\underline{8k^3 - 2k + 9}$

6. $-3n^4 + \ n^3 + 4n^2$
   $\underline{n^4 + 2n^3 - \ n^2}$

Add or subtract as indicated.

7. $(p^2 - 3p + 8) - (-2p^2 + p - 3)$ 8. $(3a + 2b - 5) + (6a + 5b - 1)$

9. $(7x^2 - 6y^2) - (3x^2 + xy - 2y^2)$ 10. $(3m - 5j + 3) - (3j - 5)$

For Questions 11–16, multiply.

11. $(a + 4)(a + 4)$ 12. $(m - 5)(m - 3)$ 13. $(2x + 3)(4x - 1)$

14. $(7k - 2)(3k + 4)$ 15. $-x(7x^2 - 4x + 1)$ 16. $4a(3a^2 - 2)$

17. **Writing** Explain why the expression $\dfrac{x^2}{3} + x$ is a quadratic binomial, but $\dfrac{3}{x^2} + x$ is not.

ANSWERS

1. _See question._
2. _See question._
3. _See question._
4. _____
5. _____
6. _____
7. _____
8. _____
9. _____
10. _____
11. _____
12. _____
13. _____
14. _____
15. _____
16. _____
17. _See question._

Test 45

(CONTINUED)

DIRECTIONS: Write the answers in the spaces provided.

Factor if possible. If not possible, write *not factorable*.

18. $9a^2 - 12a$

19. $16x^3 - 8x^2 + 4x$

20. $x^2 + 6x + 5$

21. $x^2 - 7x + 10$

22. $k^2 - 8k - 6$

23. $2x^2 + 7x - 4$

24. $2x^2 - 5x - 12$

25. $6x^2 + x - 35$

For Questions 26–31, solve each equation.

26. $x(x - 13) = 0$

27. $2y(y + 8) = 0$

28. $x^2 - 49 = 0$

29. $18x = 10x^2$

30. $(2x + 3)(4x - 5) = 0$

31. $3x^2 - 19x + 5 = 19$

32. Open-ended Problem Draw and label a rectangle formed by at least three algebra tiles. Find the perimeter and the area of your rectangle in terms of x.

ANSWERS

18. _____

19. _____

20. _____

21. _____

22. _____

23. _____

24. _____

25. _____

26. _____

27. _____

28. _____

29. _____

30. _____

31. _____

32. *See question.* _____

Test 46

TEST ON CHAPTER 10 (FORM B)

DIRECTIONS: Write the answers in the spaces provided.

ANSWERS

Tell whether each expression is a polynomial. Write *Yes* or *No*. If not, explain why not.

1. $3x^{-3} + 5x - 7$

2. $5x + 1$

3. $\frac{8}{x^2} + 7x$

Add. Give the degree of each sum.

4. $\begin{array}{r} 6m^2 + 5m - 2 \\ 2m^2 - 5m + 3 \end{array}$ 5. $\begin{array}{r} -5n^2 + 3n + 1 \\ 5n^2 - 3n + 6 \end{array}$ 6. $\begin{array}{r} -5z^4 + 8z^3 - 5z^2 \\ 3z^4 - \ z^3 + 7z^2 \end{array}$

Add or subtract as indicated.

7. $(m^2 - 2m + 3) - (-3m^2 - m + 4)$ 8. $(5p - 4q + 9) + (2p + 5q - 3)$

9. $(6t^2 - 7v^2) - (3t^2 + tv - 3v^2)$ 10. $(9x^2 + 5y - 4) - (2y - 3)$

For Questions 11–16, multiply.

11. $(b + 5)(b + 5)$ 12. $(n - 7)(n - 3)$ 13. $(3x + 2)(x - 4)$

14. $(4m - 3)(2m - 7)$ 15. $-c(3c^2 + 5c - 1)$ 16. $6a(2a^2 + 3)$

17. **Writing** Explain why the expression $\frac{x^2}{3} + x + 4$ is a quadratic trinomial, but $\frac{3}{x^2} + x + 4$ is not.

1. *See question.*

2. *See question.*

3. *See question.*

4. _____

5. _____

6. _____

7. _____

8. _____

9. _____

10. _____

11. _____

12. _____

13. _____

14. _____

15. _____

16. _____

17. *See question.*

Test 46

(CONTINUED)

DIRECTIONS: Write the answers in the spaces provided.

Factor if possible. If not possible, write *not factorable*.

18. $15a^2 - 10a$

19. $9v^3 - 12v^2 - 21v$

20. $x^2 + 10x + 9$

21. $w^2 - 5w - 5$

22. $k^2 - 8k + 7$

23. $3x^2 + 10x + 3$

24. $3x^2 + 13x - 10$

25. $6x^2 + 11x - 35$

For Questions 26–31, solve each equation.

26. $m(m - 11) = 0$

27. $4k(k + 5) = 0$

28. $x^2 - 16 = 0$

29. $16x = 6x^2$

30. $(3x - 2)(5x + 4) = 0$

31. $5x^2 + 16x + 4 = 20$

32. Open-ended Problem Draw and label a rectangle formed by at least four algebra tiles. Find the perimeter and the area of your rectangle in terms of x.

ANSWERS

18. _____

19. _____

20. _____

21. _____

22. _____

23. _____

24. _____

25. _____

26. _____

27. _____

28. _____

29. _____

30. _____

31. _____

32. *See question.* _____

NAME _____ DATE _____ SCORE _____

Test 47

DIRECTIONS: Write the answers in the spaces provided.

ANSWERS

For Questions 1 and 2, do the data show inverse variation? Write *Yes* or *No.* If they do, write an equation for *y* in terms of *x.*

1.
x	y
5	42
3	70
14	15

2.
x	y
6	50
4	75
15	20

3. For a parallelogram with fixed area, the height and the length of the base vary inversely. Suppose such a parallelogram has base length 16 in. and height 15 in. A second parallelogram with the same area has base length 48 in. Find its height.

4. In Maria's Calculus class, her score on the semester test is worth three times as much as her score on each chapter test when computing her semester average. Maria's scores on her chapter tests have been 78, 86, 79, 92, and 89. If Maria receives a score of 88 on the semester test, what is her semester average?

5. Liang is designing a theater. The owners want the theater to have 600 balcony seats and at least 250 orchestra seats. The owners plan to charge $20 per seat for balcony seats and $35 per seat for orchestra seats. How many orchestra seats must Liang include in the design so the owners receive an average of $25 per seat when the theater is full?

6. **Open-ended Problem** Write an equation that models direct variation and another equation that models inverse variation. Then graph your equations separately on the coordinate axes below.

1. _____

2. _____

3. _____

4. _____

5. _____

6. *See question.* _____

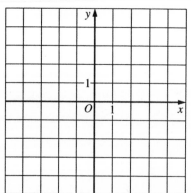
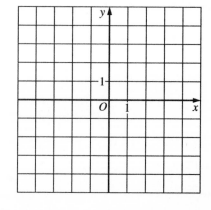

Test 48

DIRECTIONS: Write the answers in the spaces provided.

ANSWERS

Solve each equation. If the equation has no solution, write *no solution.*

1. $\dfrac{x+2}{4} = \dfrac{x-2}{6}$ **2.** $\dfrac{4}{3x} + \dfrac{5}{6} = \dfrac{-2}{x}$ **3.** $\dfrac{5}{x} + \dfrac{1}{4} = \dfrac{3}{4x}$

Solve each equation for the indicated variable.

4. $\dfrac{m+2n}{3p} = n - q$, for n **5.** $3(x + 4) = 6y^2 + 24$, for x

The formula **S = 2*lw* + 2*lh* + 2*wh*** gives the surface area of a rectangular prism with length *l*, width *w*, and height *h*. Use this information for Questions 6 and 7.

6. Solve the equation for h.

7. Find the value of h when $l = 10$ in., $w = 7$ in., and $S = 276$ in.^{2}.

8. Writing Valerie wants to graph the equation $\dfrac{x}{4} = \dfrac{y}{y+1}$ on her graphing calculator. Write out the steps that she must follow in order to enter the equation in her calculator.

1. _____

2. _____

3. _____

4. _____

5. _____

6. _____

7. _____

8. *See question.* _____

Test 49

DIRECTIONS: Write the answers in the spaces provided.

ANSWERS

For Questions 1–6, simplify each expression.

1. $\dfrac{4w}{3w-6} \cdot \dfrac{w-2}{w}$

2. $\dfrac{n-4}{10n^2} \cdot \dfrac{5n}{3n-12}$

3. $\dfrac{4x+4}{x-1} \div \dfrac{2x+2}{x}$

4. $\dfrac{7}{m+2} + \dfrac{m}{m+2}$

5. $\dfrac{5c}{c+3} - \dfrac{3}{2c+6}$

6. $\dfrac{5}{u+5} + \dfrac{25}{u(u+5)}$

7. **Writing** Explain what is meant by the phrase *simplifying a rational expression*. Give an example of a rational expression that is not simplified and show how to simplify it.

1. _____

2. _____

3. _____

4. _____

5. _____

6. _____

7. *See question.* _____

Assessment Book, ALGEBRA 1: EXPLORATIONS AND APPLICATIONS
Copyright © McDougal Littell Inc. All rights reserved.

Test 50 ....

TEST ON CHAPTER 11 **(FORM A)**

DIRECTIONS: Write the answers in the spaces provided.

ANSWERS

The math department at Pioneer High School received a $30,000 donation to be used to purchase computers. The number of computers the department can buy varies inversely with the cost per computer. Use the table for Questions 1–4.

Cost per computer	Number of computers
$2500	?
$2000	?
$1500	?
$1000	?

1. **1.** _____

2. Complete the table.

3. **2.** **Writing** Explain why this situation is an example of inverse variation.

2. *See question.* _____

3. *See question.* _____

4. _____

5. _____

6. _____

3. Write an equation showing how the number of computers they can buy varies inversely with the cost per computer.

7. _____

8. _____

4. If the math department wants to buy 25 computers, how much can the department afford to pay for each one?

For Questions 5 and 6, solve each equation for the indicated variable.

5. $5x - 2y = 16$, for y 6. $\frac{2}{k} + \frac{2}{m} = 5$, for m

7. At the furniture store where he works, Sandeep is paid a commission on his sales. He receives a commission of $10 for each table he sells. The sale of a chair earns him twice the commission for a table, while the sale of a sofa earns him three times the commission for a table. Sandeep's sales for three weeks are shown in the table at the right. How many sofas does he need to sell in the third week to earn an average of $250 in commissions per week?

	Week 1	Week 2	Week 3
Tables	5	6	2
Chairs	4	3	3
Sofas	3	4	?

8. The formula $A = \frac{1}{2}(a + b)h$ gives the area of a trapezoid with height h and bases of length a and b. Solve for a and find the value of a for which $A = 120$, $b = 10$, and $h = 6$.

Test 50

(CONTINUED)

DIRECTIONS: Write the answers in the spaces provided.

ANSWERS

Solve each equation. If the equation has no solution, write *no solution*.

9. $3 = \dfrac{2 - x}{6 + 5x}$

10. $\dfrac{5}{3} - \dfrac{x + 2}{x + 3} = \dfrac{x + 5}{3(x + 3)}$

11. $\dfrac{5}{x - 4} - \dfrac{3}{x} = \dfrac{15}{x(x - 4)}$

12. $\dfrac{3}{r - 2} - \dfrac{4}{3r} = \dfrac{9}{6(r - 2)}$

For Questions 13–21, simplify each expression.

13. $\dfrac{16xy}{5} \cdot \dfrac{10x^3}{8y^2}$

14. $\dfrac{6d}{3d + 9} \cdot \dfrac{d + 3}{7d}$

15. $\dfrac{3k + 4}{2k} \div (9k + 12)$

16. $\dfrac{x^2 - 9}{2} \div \dfrac{x + 3}{x - 3}$

17. $\dfrac{3x + 1}{2x} + \dfrac{x + 9}{2x}$

18. $\dfrac{4d + 9}{d + 5} - \dfrac{2d - 1}{d + 5}$

19. $\dfrac{x + 3}{4x^2} + \dfrac{2}{8x}$

20. $\dfrac{4}{w + 2} - \dfrac{5}{w}$

21. $\dfrac{x}{x - 5} - \dfrac{x + 5}{x}$

22. **Open-ended Problem** Write an inverse variation problem in which $k = 36$. Show a table containing at least four pairs of solution values for the inverse variation equation. Then graph the equation. Do all the values modeled by the graph make sense in your problem?

9. _____

10. _____

11. _____

12. _____

13. _____

14. _____

15. _____

16. _____

17. _____

18. _____

19. _____

20. _____

21. _____

22. *See question.* _____

Test 51

TEST ON CHAPTER 11   (FORM B)

DIRECTIONS: Write the answers in the spaces provided.

ANSWERS

The math department at Pioneer High School received a $40,000 donation to be used to purchase computers. The number of computers the department can buy varies inversely with the cost per computer. Use the table for Questions 1–4.

Cost per computer	Number of computers
$2500	?
$2000	?
$1600	?
$1000	?

1. Complete the table.

2. **Writing** Explain why this situation is an example of inverse variation.

3. Write an equation showing how the number of computers they can buy varies inversely with the cost per computer.

4. If the math department wants to buy 32 computers, how much can the department afford to pay for each one?

For Questions 5 and 6, solve each equation for the indicated variable.

5. $7x - 4y = 20$, for y

6. $\dfrac{5}{c} + \dfrac{6}{d} = 11$, for d

7. At the furniture store where she works, Sandee is paid a commission on her sales. She receives a commission of $10 for each table she sells. The sale of a chair earns her twice the commission for a table, while the sale of a sofa earns her three times the commission for a table. Sandee's sales for three weeks are shown in the table at the right. How many sofas does she need to sell in the third week to earn an average of $250 in commissions per week?

	Week 1	Week 2	Week 3
Tables	5	6	4
Chairs	4	3	5
Sofas	3	4	?

8. The formula $A = \frac{1}{2}(a + b)h$ gives the area of a trapezoid with height h and bases of length a and b. Solve for b and find the value of b for which $A = 150$, $b = 15$, and $h = 6$.

ANSWERS

1. _____

2. *See question.* _____

3. *See question.* _____

4. _____

5. _____

6. _____

7. _____

8. _____

Test 51

(CONTINUED)

DIRECTIONS: Write the answers in the spaces provided.

ANSWERS

Solve each equation. If the equation has no solution, write *no solution*.

9. $5 = \dfrac{3-x}{5+2x}$

10. $\dfrac{3}{4} - \dfrac{x+1}{x-4} = \dfrac{x-8}{4(x-4)}$

11. $\dfrac{6}{x-3} - \dfrac{2}{x} = \dfrac{30}{x(x-3)}$

12. $\dfrac{2}{r-5} - \dfrac{2}{5r} = \dfrac{6}{10(r-5)}$

For Questions 13–21, simplify each expression.

13. $\dfrac{35xy}{9} \cdot \dfrac{18y^3}{14x^2}$

14. $\dfrac{15d}{2d+10} \cdot \dfrac{d+5}{10d}$

15. $\dfrac{5c-3}{2c} \div (10c-6)$

16. $\dfrac{y^2-16}{3} \div \dfrac{y+4}{y-4}$

17. $\dfrac{5x+2}{3x} + \dfrac{x+7}{3x}$

18. $\dfrac{2c-8}{c-3} - \dfrac{1-c}{c-3}$

19. $\dfrac{x-5}{5x^2} + \dfrac{3}{10x}$

20. $\dfrac{7}{k+3} - \dfrac{2}{k}$

21. $\dfrac{n}{n-6} - \dfrac{n+6}{n}$

22. Open-ended Problem Write an inverse variation problem in which $k = 24$. Show a table containing at least four pairs of solution values for the inverse variation equation. Then graph the equation. Do all the values modeled by the graph make sense in your problem?

9. _____

10. _____

11. _____

12. _____

13. _____

14. _____

15. _____

16. _____

17. _____

18. _____

19. _____

20. _____

21. _____

22. *See question.* _____

Test 52

DIRECTIONS: Write the answers in the spaces provided.

The graph at the right models the locations where bank deposits are picked up daily by an armored car company. The driver's route is changed each day. Use this graph for Questions 1–3.

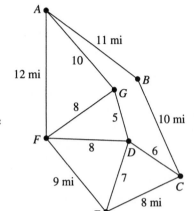

1. **Open-ended Problem** Find a short path for one driver to use to visit each of the locations.

2. **Open-ended Problem** Find a different path for the driver to use the next day.

3. Suppose two armored cars will be used. Each driver will pick up some of the bank deposits. Suggest a fair division of the locations. Explain why your division is fair.

4. **Writing** Write an algorithm for multiplying two binomials.

5. **Writing** Explain the meaning of the term *greedy algorithm*.

ANSWERS

1. *See question.* _____

2. *See question.* _____

3. *See question.* _____

4. *See question.* _____

5. *See question.* _____

Test 54 ························

TEST ON CHAPTER 12   **(FORM A)**

DIRECTIONS: Write the answers in the spaces provided.

1. Write an algorithm to divide two proper fractions.

ANSWERS

1. *See question.* _____

2. _____

3. *See question.* _____

4. *See question.* _____

5. *See question.* _____

For Questions 2–4, use this graph.

2. Find the shortest path from vertex *W* to vertex *Z*.

3. Draw the shortest tree for the graph.

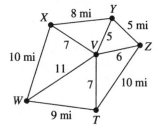

4. Start at vertex *Z*. Find a short path through all the points.

5. **Open-ended Problem** Create a graph with at least seven points. Label the vertices of your graph and include the distances between them. Write two questions based on your graph, then answer the questions.

Test 54

(CONTINUED)

DIRECTIONS: **Write the answers in the spaces provided.**

6. Write a cut-and-choose algorithm for this situation: Cheryl and Ron each want to eat the last apple in the refrigerator.

7. In how many ways can a five-member board of directors be chosen from among nine persons?

In a student body election, there are three candidates for president, four candidates for vice-president, and five candidates for secretary.

8. How many possible groups of officers are there?

9. If two of the candidates for each office are females, in how many ways can there be an all-female set of officers? an all-male set of officers?

Evaluate.

10. $_7C_5$ **11.** $_{10}C_2$ **12.** $_6P_3$ **13.** $_8P_4$

For Questions 14 and 15, four coins are tossed.

14. Find the probability of getting exactly 4 heads.

15. Find the probability of getting exactly 1 head.

16. Two dice are rolled. What is the probability that at least one of the dice will show a number less than 4?

17. A special deck of cards contains three each of the numbers from 1 to 8 and four each of the numbers 9 and 10. One card is drawn at random from the deck. What is the probability that the card is a number greater than 7?

18. **Writing** Explain the meaning of a probability of zero and a probability of one.

ANSWERS

6. *See question.*

7. _____

8. _____

9. _____

10. _____

11. _____

12. _____

13. _____

14. _____

15. _____

16. _____

17. _____

18. *See question.*

Test 55

TEST ON CHAPTER 12 (FORM B)

DIRECTIONS: Write the answers in the spaces provided.

1. Write an algorithm to multiply two mixed numbers.

ANSWERS

1. *See question.* _____

2. _____

3. *See question.* _____

4. *See question.* _____

5. *See question.* _____

For Questions 2–4, use this graph.

2. Find the shortest path from vertex W to vertex Z.

3. Draw the shortest tree for the graph.

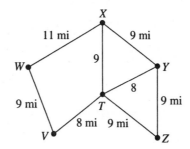

4. Start at vertex Y. Find a short path through all the points.

5. **Open-ended Problem** Create a graph with at least five points. Label the vertices of your graph and include the distances between them. Write two questions based on your graph, then answer the questions.

Test 55

(CONTINUED)

DIRECTIONS: Write the answers in the spaces provided.

ANSWERS

6. Write a cut-and-choose algorithm for this situation: Jane and Eli each want the car this evening.

6. *See question.* _____

7. _____

8. _____

9. _____

7. In how many ways can a six-member board of directors be chosen from among eleven persons?

10. _____

In a student body election, there are three candidates for president, four candidates for vice-president, and three candidates for secretary.

11. _____

8. How many possible groups of officers are there?

9. If two of the candidates for each office are seniors, in how many ways can there be an all-senior set of officers? a non-senior set of officers?

12. _____

13. _____

Evaluate.

10. $_6C_3$ 11. $_8C_4$ 12. $_7P_5$ 13. $_{10}P_2$

14. _____

For Questions 14 and 15, four coins are tossed.

14. Find the probability of getting exactly 3 tails.

15. _____

15. Find the probability of getting exactly 0 tails.

16. _____

16. Two dice are rolled. What is the probability that at least one of the dice will show a number greater than 4?

17. _____

17. A special deck of cards contains two each of the odd numbers from 1 to 8 and three each of the even numbers from 1 to 8. One card is drawn at random from the deck. What is the probability that the card is a number greater than 5?

18. *See question.* _____

18. **Writing** Explain the meaning of the mathematical term "factorial." Give an example to illustrate your explanation.

NAME _____ DATE _____ SCORE _____

Test 56

DIRECTIONS: Write the answers in the spaces provided.

ANSWERS

Give the horizontal intercept and the vertical intercept of each equation. Then graph each equation.

1. $2x + 5y = 10$

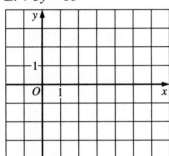

2. $3x - 5y = 30$

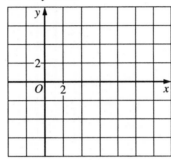

For Questions 3–6, solve each system of equations.

3. $3x + 4y = 23$
$5x + y = 27$

4. $4x - 2y = -10$
$3x - 2y = -9$

5. $7x - 2y = -1$
$-6x + 3y = 6$

6. $7x + 6y = 20$
$3x - 2y = 4$

7. At a high school musical, 400 tickets were sold. Adult tickets sold for $8 and student tickets sold for $5. If the total amount collected was $2450, how many adult tickets were sold?

Match each equation with *all* the terms below that correctly describe the equation.

A. linear   **B. nonlinear**   **C. quadratic**   **D. absolute value**

8. $y = 4x^2 - 3$ **9.** $y = -|3x| + 1$ **10.** $y = -4x + 5$

Solve each equation algebraically.

11. $x^2 + 29 = 45$ **12.** $3x^2 - 4 = 143$ **13.** $\frac{2}{3}x^2 + 6 = 30$

Solve each equation using the quadratic formula.

14. $x^2 + 10x = -7$ **15.** $3x^2 = 2x + 3$ **16.** $7x + 1 = 4x^2$

Find the number of solutions of each equation.

17. $3x^2 + 3 = -6x$ **18.** $4x^2 = x - 5$ **19.** $3x = 1 + 2x^2$

1. _____
2. _____
3. _____
4. _____
5. _____
6. _____
7. _____
8. _____
9. _____
10. _____
11. _____
12. _____
13. _____
14. _____
15. _____
16. _____
17. _____
18. _____
19. _____

Test 56

(CONTINUED)

DIRECTIONS: Write the answers in the spaces provided.

ANSWERS

Graph each inequality or system of inequalities.

20. $y > x - 2$

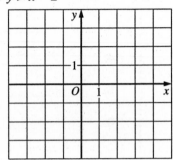

21. $x - 2y \leq 8$

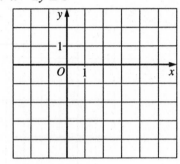

22. $x \geq -2$
   $y \leq -x$

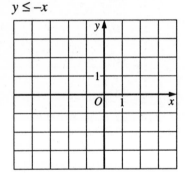

23. $2x + 5y > 10$
   $x - y < 3$

20. *See question.* _____

21. *See question.* _____

22. *See question.* _____

23. *See question.* _____

24. _____

25. _____

26. _____

27. _____

28. _____

29. _____

30. _____

31. _____

32. _____

33. _____

Evaluate each expression when $x = -4$ and $y = 2$.

24. $\dfrac{x}{y^2} + (3 - y)^2$

25. $\dfrac{-3x^2 + 2y^3}{xy}$

26. $3x^2y^{-3}$

27. x^y

28. xy^0

A new house built in 1995 was valued at \$150,000. Each year the value of the house increases by 2%.

29. Write an equation giving the value y of the house x years after 1995.

30. How much will the house be worth in the year 2010?

Simplify each expression.

31. $(2x^3)^6$

32. $\dfrac{x^{-3}y^4}{y^3z^{-5}}$

33. $\left(\dfrac{-3a^{-3}b^7}{2a^4b^0}\right)^3$

Test 56

(CONTINUED)

DIRECTIONS: Write the answers in the spaces provided.

Add, subtract, or multiply as indicated.

34. $(w^2 - 5w + 6) - (-3w^2 + 2w + 7)$ **35.** $(m + 8)^2$

36. $(5c + 3d - 4) + (-3c - 6d + 2)$ **37.** $(p - 12)(2p + 3)$

Factor if possible. If not, write *not factorable*.

38. $v^2 - 10v + 25$ **39.** $r^2 - 12r - 45$ **40.** $6m^2 - 19m - 7$

For Questions 41–43, use this graph.

41. Find the shortest path from vertex E to vertex A.

42. Start at vertex B. Find a short path through all the points.

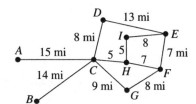

43. Open-ended Problem Change the graph by removing vertices and paths so that the shortest tree includes exactly 5 paths. Label all vertices in your tree.

In a freshman class election, there are four candidates for president, four candidates for vice-president, and three candidates for secretary.

44. How many possible groups of officers are there?

45. If two of the candidates for each office are female, in how many ways can there be an all-female set of officers? an all-male set of officers?

Two dice are rolled.

46. Find the probability of rolling a sum of 3.

47. Find the probability of rolling a sum less than or equal to 5.

48. What is the probability that at least one of the dice will show a number greater than or equal to 5?

ANSWERS

34. _____

35. _____

36. _____

37. _____

38. _____

39. _____

40. _____

41. _____

42. *See question.* _____

43. *See question.* _____

44. _____

45. _____

46. _____

47. _____

48. _____

Assessment Book, ALGEBRA 1: EXPLORATIONS AND APPLICATIONS

85

Test 56

(CONTINUED)

DIRECTIONS: Write the answers in the spaces provided.

ANSWERS

Simplify each expression.

49. $\dfrac{8d}{4d + 12} \cdot \dfrac{d + 3}{2d}$ **50.** $\dfrac{5k - 3}{2k} \div \dfrac{15k - 9}{8}$ **51.** $\dfrac{3x}{2x - 5} - \dfrac{x + 1}{2x + 5}$

For Questions 52–57, solve each equation. If the equation has no solution, write _no solution_.

52. $3z(4z - 5) = 0$ **53.** $9x^2 - 4 = 0$ **54.** $24x = 9x^2$

55. $6x^2 + x = 12$ **56.** $5 = \dfrac{x + 4}{2x - 1}$ **57.** $\dfrac{x + 3}{x - 2} = \dfrac{x + 2}{x - 2} + \dfrac{1}{2x}$

58. Solve the equation $\dfrac{3}{p} + \dfrac{5}{m} = 2$ for m.

59. The formula $V = \dfrac{1}{3}lwh$ gives the volume of a pyramid with height h and rectangular base of length l and width w. Solve for h and then find the value of h for which $V = 5000$ in.^{3}, $l = 25$ in., and $w = 15$ in.

60. Writing Use the equations $y = 2x$ and $xy = 2$. Identify which equation represents inverse variation and which represents direct variation. Graph each equation. Then describe at least two distinctive characteristics of the graphs of each type of variation.

49. _____

50. _____

51. _____

52. _____

53. _____

54. _____

55. _____

56. _____

57. _____

58. _____

59. _____

60. _See question._

Assessment Book, ALGEBRA 1: EXPLORATIONS AND APPLICATIONS

Test 57

CUMULATIVE TEST ON CHAPTERS 1–12

DIRECTIONS: Write the answers in the spaces provided.

1. Find the mean, the median, and the mode(s) of the scores below.
   90, 86, 94, 85, 79, 86, 75, 87, 68, 93

Simplify each expression.

2. $-14 - (-31)$ 3. $(-7)(-4)$ 4. $-8 + |-17|$

For Questions 5 and 6, use matrices A and B.

$$A = \begin{bmatrix} 9 & 13 & -7 & 0 \\ -5 & -4 & 15 & 1 \end{bmatrix} \qquad B = \begin{bmatrix} 0 & -3 & 3 & -5 \\ 10 & -1 & -2 & 4 \end{bmatrix}$$

5. What are the dimensions of each matrix?

6. Add the two matrices.

7. If $k < m$, which of the following inequalities is true?

   A. $\dfrac{k}{3} > \dfrac{m}{3}$ B. $-8k > -8m$ C. $\dfrac{11k}{13} > \dfrac{11m}{13}$

Solve each equation. If an equation is an *identity* or if there is *no solution*, say so.

8. $\dfrac{x}{5} - 2 = 2$ 9. $\dfrac{9}{d} = \dfrac{6}{d+3}$ 10. $\dfrac{5}{6}v - \dfrac{1}{3} = \dfrac{3}{4}v + \dfrac{1}{6}$

11. $7k = 19k + 6$ 12. $2.4d - 7.5 = 8.1$ 13. $18 + 6m = -4m - 7$

14. $x^2 + 16 = 52$ 15. $2x^2 - 5x = 3$ 16. $x^2 + 10x - 39 = 0$

Solve each proportion.

17. $\dfrac{3.4}{4.4} = \dfrac{x-2}{x}$ 18. $\dfrac{24}{8} = \dfrac{3x+6}{4x-1}$ 19. $\dfrac{x+5}{x-3} = \dfrac{x+4}{x-3} + \dfrac{1}{2x}$

ANSWERS

1. _See question._ _____

2. _____

3. _____

4. _____

5. _____

6. _See question._ _____

7. _____

8. _____

9. _____

10. _____

11. _____

12. _____

13. _____

14. _____

15. _____

16. _____

17. _____

18. _____

19. _____

Test 57

(CONTINUED)

DIRECTIONS: Write the answers in the spaces provided.

Solve each inequality.

20. $-7y + 9 > 51$

21. $5(x - 4) < 6x + 8 + x$

Find an equation of the line through each pair of points.

22. $(6, -2), (3, 4)$ **23.** $(-3, -1), (4, -1)$ **24.** $(3, -1), (3, 6)$

Graph each equation or inequality on a coordinate plane.

25. $y = -\frac{2}{3}x + 3$

26. $-3x + 2y \geq 4$

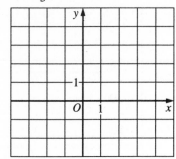

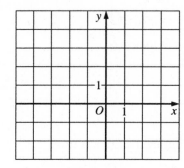

Tell whether the given lengths can be the sides of a right triangle.

27. $4, 5, 6$ **28.** $21, 28, 35$ **29.** $10, 16, 2\sqrt{89}$

Solve each equation algebraically or using the quadratic formula.

30. $x^2 = 77 + 4x$ **31.** $2x^2 + 3x = 20$ **32.** $3x^2 + 2x = 5$

For Questions 33 and 34, find each product.

33. $(2 + 3\sqrt{8})(4 + 5\sqrt{2})$

34. $(8 + \sqrt{5})(8 - \sqrt{5})$

35. Writing Explain the difference between geometric probability and experimental probability.

20. _____

21. _____

22. _____

23. _____

24. _____

25. *See question.*

26. *See question.*

27. _____

28. _____

29. _____

30. _____

31. _____

32. _____

33. _____

34. _____

35. *See question.*

Test 57 ·······························

(CONTINUED)

DIRECTIONS: Write the answers in the spaces provided.

36. **Open-ended Problem** Using the variable x, write binomial expressions for the length and the width of a rectangle. Draw and label your rectangle. Then write variable expressions for its perimeter and area.

Graph each system of inequalities.

37. $x \geq -2$
    $y \leq -0.5x$

38. $2x - 3y < 9$
    $x + y < 2$

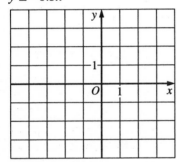 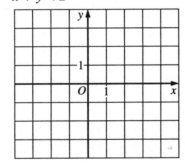

The mass y in grams of a colony of bacteria x hours after 6 A.M., July 28, is modeled by the equation $y = 2.9(1.68)^x$. Estimate the mass at each time.

39. 10 A.M., July 28

40. 2 A.M., July 28

Solve each system of equations.

41. $2x + 4y = 26$
    $x + y = 8$

42. $3x + 2y = 10$
    $-7x + 2y = 30$

43. $4x - 3y = 14$
    $3y = 4 - 5x$

Factor if possible. If not, write *not factorable*.

44. $6j^2 - 3jkm$

45. $v^2 - 18v + 36$

46. $6x^2 + x - 15$

Simplify each expression.

47. $\dfrac{k^4 r^2}{k^{-4} r^5}$

48. $\left(3x^3\right)^4$

49. $\left(\dfrac{-2a^{-3}b^5}{4a^3b^2}\right)^3$

ANSWERS

36. *See question.* _____

37. *See question.* _____

38. *See question.* _____

39. _____

40. _____

41. _____

42. _____

43. _____

44. _____

45. _____

46. _____

47. _____

48. _____

49. _____

Test 57

(CONTINUED)

DIRECTIONS: Write the answers in the spaces provided.

For Questions 50–53, add, subtract, or multiply as indicated.

50. $(w^2 - 8w + 2) - (-5w^2 + 7w + 3)$

51. $(p - 9)(2p + 7)$

52. $(6c + 4d - 2) + (-3c - 8d + 2)$

53. $(2x^2 - 5)(2x^2 - x + 5)$

54. Solve the equation $\dfrac{7}{x} + \dfrac{3}{m} = 5$ for m.

Simplify each expression.

55. $\dfrac{10r}{5r + 15} \cdot \dfrac{3r + 9}{5r}$

56. $\dfrac{4d - 7}{3d} \div \dfrac{8d - 14}{9}$

57. $\dfrac{x - 2}{3x + 5} - \dfrac{2x}{3x - 5}$

58. $\dfrac{6n + 1}{4n^2 - 8n} + \dfrac{5}{3(n - 2)}$

For Questions 59 and 60, use the graph at the right.

59. Find the shortest path from vertex E to vertex A.

60. Start at vertex B. Find a short path through all the points.

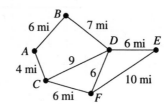

61. In how many ways can a seven-member board of directors be chosen from among twelve cabinet members?

62. Two dice are rolled. Find the probability that exactly one of the dice will show a number less than 5.

63. **Writing** Describe how the graph of the equation $y = -0.8x^2$ compares to the graph of $y = x^2$.

ANSWERS

50. _____

51. _____

52. _____

53. _____

54. _____

55. _____

56. _____

57. _____

58. _____

59. _____

60. *See question.*

61. _____

62. _____

63. *See question.*

Chapter 1

1. **Open-ended Problem** What is a variable? Include examples with your explanation.

2. **Project** Cut out articles from newspapers and magazines that contain examples of inequalities. Write a mathematical inequality that could represent each example.

3. **Project** Write a play-by-play commentary for part of a football game. Next to each play write a mathematical statement describing the progress of the team. Use positive and negative integers in your description.

4. **Group Activity** Work in a group of four students. In your group, test different types of numbers in the two statements below. As a group, write a general description to answer each of the questions.

   a. Which values for a and b make the statement $|a| + |b| = |a + b|$ true? Which values for a and b make the statement false?

   b. Which values for a and b make the statement $|a| - |b| = |a - b|$ true? Which values for a and b make the statement false?

5. **Performance Task** A set of quiz grades for two different classes is given in the table at the right.

Class 1	Class 2
5	8
7	9
10	8
10	6
8	8
7	10
5	8
10	7
8	9
10	7

   a. Calculate the mean for each class. What does the mean tell you about each class?

   b. Draw a histogram for each class. What does the histogram tell you about each class?

   c. Which class do you think would be easier for a teacher to teach? Explain why you think so.

   d. Why should a teacher be cautious about making assumptions based on the mean of a set of quiz grades?

6. **Research Project** Look up the word *average* in the dictionary and copy its definition. If you can find the definition of *mean* embedded in your definition, underline it. If you can find the definition of *mode* embedded in your definition, circle it. If you can find the definition of *median* embedded in your definition, put a box around it. What does this tell you about the word "average?"

7. **Open-ended Problem** Matrix A below represents the inventory of a music store at the beginning of a month. Matrix B represents the merchandise delivered early in that month.

$$A = \begin{array}{c} \\ \text{Tapes} \\ \text{CDs} \end{array} \overset{\text{Classical \ Rock}}{\begin{bmatrix} 900 & 1100 \\ 1500 & 1800 \end{bmatrix}} \quad B = \begin{array}{c} \\ \text{Classical} \\ \text{Rock} \end{array} \overset{\text{Tapes \ CDs}}{\begin{bmatrix} 250 & 400 \\ 700 & 650 \end{bmatrix}}$$

   a. Explain why it does not make sense to add matrix A to matrix B to get a total inventory after the delivery. Rewrite the matrices so that addition is possible.

   b. Add the rewritten matrices from part (a). Explain what you did to add them and what the resulting matrix represents.

8. **Performance Task** Make an integer timeline of your life. Your birth date is represented by 0 on the timeline. Use positive and negative integers to represent the years. Events before your birth will be located along the negative part of the timeline. Mark and label the locations for at least three family events that occurred before your birth. Include photos or drawings of at least five important personal events after your birth. Have a parent or relative help you choose the events and pictures. *Extension:* Use fractions or decimals to represent your events to the nearest day or month.

9. a. Draw a geometric figure that has a perimeter of $2x + 3y$. Label the sides of your figure.

   b. Draw a quadrilateral that has a perimeter of $2x + 3y$. Label the sides of your figure.

   c. Draw a rectangle that has a perimeter of $2x + 3y$. Label the sides of your figure.

Chapter 2

1. **Project** Plumber A charges $30 per hour plus a $20 service fee per job. Plumber B charges $25 per hour plus a $40 service fee per job. Use tables and graphs to compare the plumbers' rates. Assuming the quality of their work is similar, which plumber's rates are better? Explain your answer.

2. **Open-ended Problem** Use the graph below to describe each student's school performance. Explain the relationship between attendance and grades for each student.

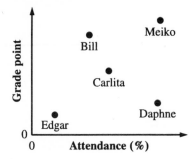

3. **Open-ended Problem** What are some questions that could be asked based on the graph shown below? What are the answers to those questions?

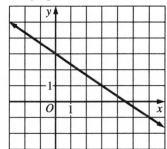

4. **Performance Task** Find an example of a function in your life. Describe the function. Describe the domain and range. Collect some data from your function. Represent your data in a graph or table. Write an equation that can be used to describe your function.

5. Mathematicians call the variable that represents the values in the domain of a function the *independent* variable and the variable that represents the values in the range of a function the *dependent* variable. Choose a real-life example of a function that is modeled by an equation from this chapter. Use the function to discuss why the terms *independent variable* and *dependent variable* might be used.

---

6. **Group Activity** There is a formula in physics called Hooke's Law. This law defines the relationship between the mass of an object and the amount that a spring will stretch when the object is suspended from it. To study this relationship, follow the directions below.

   a. Gather the following materials: 15 to 20 pennies, paper cup with a wire handle, a loose spring, 2 meter sticks.

   b. Suspend one meter stick across two chairs.

   c. Form a hook on one end of the spring.

   d. Attach the spring to the meter stick.

   e. Attach the cup to the other end of the spring.

   f. Measure and record the length of the spring.

   g. Place one or more pennies in the cup. Measure and record the new length of the spring.

   h. Repeat this procedure at least 10 times using a different number of pennies for each trial. Record the number of pennies and the length of the spring for each trial.

   i. Graph your results. What is your domain? What is your range? Justify your choices for domain and range. What other information can you get from your graph?

   j. Repeat the experiment, measuring the distance from the bottom of the cup to the floor. How does this change affect the graph?

   k. If you repeated the experiment with another mass (like dried beans), how do you think the graph would change?

7. **Project** Create a comic strip about solving equations. In your comic strip, give students at least one tip to help them avoid making common mistakes when solving an equation.

8. **Open-ended Problem** Make up a story to go with the function graph shown below.

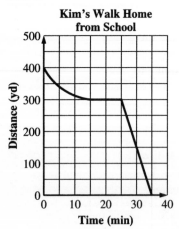

Kim's Walk Home from School

Assessment Book, ALGEBRA 1: EXPLORATIONS AND APPLICATIONS

Chapter 3 ·······································

ALTERNATIVE ASSESSMENT

1. a. Some people think of ski slopes when they read about the mathematical term *slope*. The drawings below represent three different "ski" slopes.

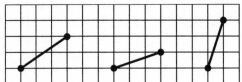

If you were rating these ski slopes in terms of difficulty, which would get the highest rating? Use the grids shown to estimate the numerical slope of each figure. Discuss how the numerical slopes confirm your ratings.

b. Find the numerical slope of a horizontal line. What would be an appropriate ski rating for a "hill" with a similar slope? Discuss how you might use the "ski" slope to help remember the numerical slope.

c. Find the numerical slope of a vertical line. What would be an appropriate ski rating for a "hill" with a similar slope? Discuss how you might use the "ski" slope to help remember the numerical slope.

2. Open-ended Problem William needed to calculate the slope of a line containing the points (1, 4) and (–2, –5). He wrote the following calculation: $\frac{4 - (-5)}{-2 - 1} = \frac{9}{-3} = -3$. Then William drew the graph of the line and realized the slope of the line containing (1, 4) and (–2, –5) should be positive. Find and discuss William's error. What can William do to avoid repeating this error?

3. Group Activity Discuss the following questions in your group.

a. Find a real-life function that has a graph with a positive slope. Graph the function.

b. Explain how you knew that the slope of the line would be positive.

c. Can a function whose graph has a positive slope be called an *increasing* function? Explain.

d. Find a real-life function that has a graph with a negative slope. Graph the function.

e. Explain how you knew that the slope of the line would be negative.

f. Can a function whose graph has a negative slope be called an *decreasing* function? Explain.

---

4. Draw a line of fit for each scatter plot. Estimate the equation for your line of fit.

a.

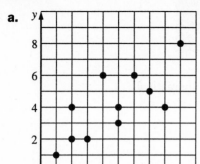

b.

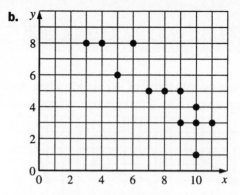

c.

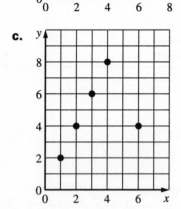

d.

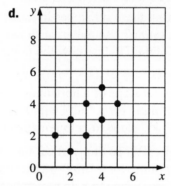

5. Performance Task Suppose a line contains the points $(0, 0)$ and (q, r).

   a. What happens to the slope of the line if q and r are both tripled?

   b. Suppose the horizontal and vertical distances from the origin to (q, r) were originally measured in inches. What happens to the slope of the line if these distances are measured again in centimeters? Make a generalization about the effect of the unit of measurement on the slope of a line.

6. a. Find two points on a graph that models a rate of change of $\frac{3}{4}$.

   b. Is there more than one correct answer to part (a)? If so, find other examples of pairs of points for which the rate of change is $\frac{3}{4}$.

   c. Is there a generalization that can be made about all the pairs of points that have a rate of change of $\frac{3}{4}$?

7. Open-ended Problem What are some questions that could be asked based on the graph shown at the right? What are the answers to your questions?

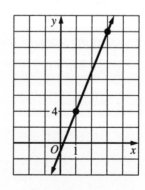

Chapter 4 ···

1. **a.** Solve the inequality $4x + 2 \leq 2(x - 1)$. Graph the solution on a number line.

   b. Explain how to check your answer.

   c. Name three values that would be appropriate to use when checking your solution.

   d. Compare the procedure for checking the answer to an inequality with the procedure for checking the answer to an equation.

2. **a.** Write an inequality that has no solutions.

   b. Write an inequality for which all numbers are solutions.

3. **a.** Create a table of x- and y-values for the function $y = 3x - 1$. Create a table of x- and y-values for the function $y = -2x + 24$.

   b. Use the two tables from part (a) to discuss the solution of the equation $3x - 1 = -2x + 24$.

   c. Use the two tables from part (a) to discuss the solutions of the inequality $3x - 1 \leq -2x + 24$.

4. **Open-ended Problem** Write three other inequalities that are equivalent to $x > -5$.

5. **Project** Call a nearby amusement park or skating rink that offers a season pass. Using a graph or table, compare the cost of the season pass to the single admission charge. Write an article for the school newspaper discussing the advantages and disadvantages of buying a season pass. In your article, include the number of visits you would need to make in order for the season pass to be cost-effective.

6. **Group Activity** Use tables to solve the equation $4x + \frac{1}{3} = 3x + 1$. (*Hint:* If the solution is not immediately evident, sometimes a graph will help.) What decisions did your group make in trying to find a solution? How can your group be assured that the solution is correct? How do you know that there is only one solution?

---

5. **Open-ended Problem** Solve the proportion $\frac{x}{15} = \frac{10}{6}$ without using the means-extremes property. (*Hint:* Use what you know about solving equations.)

6. **Performance Task** Make a scale model of your favorite room in your home. Be sure to indicate the scale of your model.

7. **Research Project** Compare the size of a doll or action figure to an average-sized adult. Consider the lengths of the arms and the legs, the circumference of the head, and so on. What scale was used to reduce the doll or action figure from the size of an average adult? Are the proportions accurate? The Golden Rectangle was used by many ancient artists to keep the images of human bodies in proportion. Investigate and report on the Golden Rectangle.

1. a. Name two irrational numbers between 9 and 10.

   b. Name two rational numbers between 9 and 10.

2. Open-ended Problem The following question appeared on a quiz. Tyler's work is shown to the right of the figure.

Find the length of $\overline{CD}$.

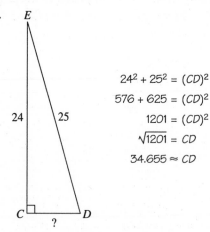

$$24^2 + 25^2 = (CD)^2$$
$$576 + 625 = (CD)^2$$
$$1201 = (CD)^2$$
$$\sqrt{1201} = CD$$
$$34.655 \approx CD$$

As you can see from his work, Tyler found the length of the missing side of this triangle to be $\sqrt{1201}$. He knows something is wrong because $\sqrt{1201} \approx 34.66$ and the length of $\overline{CD}$ appears to be much less than 35 units. Find the error in Tyler's work. Write a paragraph explaining his error. Think of a helpful hint to give Tyler so that he can remember to avoid this error next time.

3. a. Find several values of a and b such that $\sqrt{a^2 + b^2} = a + b$.

   b. Find several values of a and b such that $\sqrt{a^2 + b^2} \neq a + b$.

   c. Explain why the statement in part (a) is not always true.

4. Performance Task Describe several ways that you could verify the equality $\sqrt{18} + \sqrt{50} = 8\sqrt{2}$.

5. Research Project Look up the definitions for *scalene triangle*, *isosceles triangle*, and *equilateral triangle* in a dictionary. Write the definitions in your own words. Using graph paper, draw (if possible) a scalene *right* triangle, an isosceles *right* triangle, and an equilateral *right* triangle.

6. Open-ended Problem Describe the method you would use to put the following values in order.
$$\sqrt{2} - \sqrt{3},\ \sqrt{2} + \sqrt{3},\ \sqrt{2} + 3,\ \sqrt{2} - 3,\ 2 + \sqrt{3}$$

7. Project A lifeguard must be able to quickly swim the longest distance in a swimming pool.

    a. Visit a rectangular pool in or near your community. Describe the longest distance that a lifeguard might have to swim.

    b. Measure the length and width of the swimming pool.

    c. Calculate the diagonal length of the swimming pool.

    d. Ask the lifeguard how many seconds it would take her or him to swim the longest distance in the swimming pool.

8. Performance Task Follow the directions below to verify the Pythagorean theorem.

    1. Draw a right triangle.

    2. Measure one side of the triangle.

    3. On a sheet of heavy paper (like tagboard), draw a square that has a side length equal to the length you measured in Step 2.

    4. Repeat Steps 2 and 3 for the other two sides of the triangle you drew.

    5. Cut out your three squares and place the second largest square on the largest square. Now cut the smallest square into pieces that can be arranged so that the largest square is completely covered. (*Note:* There is more than one way to do this.)

Explain how this procedure verifies the Pythagorean theorem for your triangle. Does this example prove that the Pythagorean theorem is always true?

1. **Open-ended Problem** Write two equations whose graphs intersect at the point (1, 2).

2. **Open-ended Problem** Use the graph shown below to find the slope-intercept form of the equation $2x + 4y = 8$.

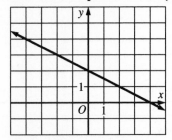

3. **Open-ended Problem** Find the equations of three lines that are parallel. Write the equations in both slope-intercept form and in standard form.

4. **a.** Write a system of two equations in standard form that has one solution.

   b. Write a system of two equations in standard form that has no solutions.

   c. Write a system of two equations in standard form that has an infinite number of solutions.

5. **Open-ended Problem** Write a system of four inequalities whose solution region is a trapezoid.

6. **a.** Find the vertices of the triangular solution region formed by the following system.

   $$y \le \frac{1}{5}x + 2$$
   $$y \ge x - 2$$
   $$x \ge 0$$

   b. Find three ordered pairs that are solutions of the system given in part (a).

7. **a.** Describe and sketch several possible solution regions for a system of two linear inequalities.

   b. Describe and sketch several possible solution regions for a system of three linear inequalities.

1. **Open-ended Problem** What do you think is the maximum number of solutions of **(a)** a linear equation, **(b)** a quadratic equation, and **(c)** a cubic equation? Explain why.

2. **Open-ended Problem** Estimate the equations for each of the parabolas shown in the graph at the right. Use your graphing calculator to check your equations.

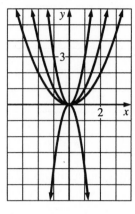

3. **Group Activity** Work in groups of four students. (This activity can also be done with the whole class.) One member of the group is chosen as the "equation reader," who reads off a list of related equations. The rest of the group members demonstrate the shape of the graph of each equation with their arms. The equation reader checks for the accuracy of the "arm graphs." The equation reader should use a list of equations like the ones given below. (Notice that the group members return to the basic graph after the graph of each related equation.)

1. $y = x^2$ ← This is the basic graph.
   $y = 2x^2$
   $y = x^2$
   $y = 3x^2$
   $y = x^2$
   $y = 0.5x^2$
   $y = x^2$
   $y = -x^2$

2. $y = x$ ← This is the basic graph.
   $y = 2x$
   $y = x$
   $y = 3x$
   $y = x$
   $y = 0.5x$
   $y = x$
   $y = -x$

Extension: Equations such as $y = 4x^2 + 3$ and $y = 4x + 3$ might also be used.

4. a. Find an example of a real-life linear function that can be represented by an equation. Write the equation and describe the domain and range of the function.

    b. Find an example of a real-life nonlinear function that can be represented by an equation. Write the equation and describe the domain and range of the function.

5. a. Complete the tables below for the quadratic equation $y = x^2 - x - 20$.

x	y
−6	?
−5	?
−4	?
−3	?
−2	?
−1	?
0	?

x	y
1	?
2	?
3	?
4	?
5	?
6	?

    Use your tables to find the solutions of the equation $0 = x^2 - x - 20$. How do you identify the solutions in the tables?

    b. Generate your own table of values for the equation $y = x^2 + 6x - 7$. Use your table to find the solutions of the equation $0 = x^2 + 6x - 7$.

    c. Generate your own table of values for the equation $y = 2x^2 + 3x + 1$. Use your table to find the solutions of the equation $0 = 2x^2 + 3x + 1$. (*Hint:* Using integer values in your table will only reveal one of the solutions.)

6. Project Use the quadratic formula to solve the equation $0 = x^2 - 2x - 5$. Describe how you can use the graph of the equation $y = x^2 - 2x - 5$ to verify your solutions to the equation $0 = x^2 - 2x - 5$. Sketch this graph and verify your solutions.

Chapter 9 ·····················

1. **a.** Compare $\left(\frac{3}{4}\right)^2$ to $\frac{3^2}{4}$.

   b. Compare $(-3)^2$ to -3^2.

   c. Compare $(5x)^2$ to $5x^2$.

2. **Open-ended Problem** Use the rules of exponents to prove that $(27)^2 = (9)^3$.

3. **Open-ended Problem** Some students confuse $x^2 \cdot x^3$ with $(x^2)^3$. Think of a way to help students remember that $x^2 \cdot x^3 = x^5$ and $(x^2)^3 = x^6$.

4. **a.** If you double the length of all the edges of a cube, how do the surface area and the volume of the cube change?

   b. If you double the length of the four vertical edges of a cube, how do the surface area and the volume of the cube change?

5. **Performance Task** Describe a real-life situation that can be represented by the equation $y = 250(1.08^x)$. Make a table of ten values for x and y. Explain what these x- and y-values mean in your situation.

6. **Open-ended Problem** Compare and contrast the linear function $y = 3x + 2$ with the exponential function $y = 2(3^x)$.

7. **Performance Task** Identify each of the following functions as a model of *exponential growth* or *exponential decay*.

   a. The value of a certain new car depreciates by 18% every year for the first 10 years.

   b. The half-life of carbon-14 is 5730 years.

   c. The cost of living in a certain city has increased by an average of 2.5% per year for the last 5 years.

   d. If has been predicted that the amount of information will be doubling every 73 days by the year 2020.

   e. The population of a bacteria colony is doubling every 40 min.

1. a. Give an example of a linear monomial, a quadratic trinomial, and a cubic binomial.

   b. Explain the difference between classifying a polynomial by its degree and classifying a polynomial by its number of terms.

2. a. Draw a tile model that represents the product $(x + 2)(x + 4)$ as an area. Then represent the area as a polynomial in standard form.

   b. Use tiles to factor the following expressions. Describe your procedure.

   $x^2 + 5x + 4$ $2x^2 + 7x + 6$ $x^2 + 2x - 3$

3. a. Make up three polynomials that each have a factor of $x - 3$.

   b. Use each polynomial from part (a) to write an equation in the form $y = \underline{\ ?\ }$.

   c. What would be true about the graphs of all three equations you wrote in part (b)? Explain your conclusions.

4. Performance Task Explain how you decide which method (factoring, graphing, or the quadratic formula) you will use to solve a quadratic equation.

5. a. How could you use graphing to prove that $(x + 3)(x - 4)$ is the factored form of $x^2 - x - 12$?

   b. Is $(3x + 1)(-x - 5)$ the factored form of $-3x^2 + 16x - 5$? Use graphing to check.

6. a. How is the degree of the sum of two polynomials related to the degrees of the two polynomials being added?

   b. How is the degree of the product of two polynomials related to the degrees of the two polynomials being multiplied?

7. Project Suppose you have been hired to design a rectangular box for a new brand of cereal. You are to use a 16 in. by 20 in. piece of cardboard to do the job. You need to design your box so that it will hold the greatest possible amount. *Note:* Do not include the lid in your design.

a. Cut a square out of each corner. Fold up the sides to form a box.

b. Find the surface area and volume of your box.

c. Repeat parts (a) and (b) for two more boxes.

d. Let x = the length of a side of the square cut from each corner. Write an equation to represent the volume of the box in terms of x.

e. Graph the equation with a graphing calculator. Find the maximum volume. What happens when $x > 8$?

f. Why is the shape of the box with the maximum volume not a cube?

g. If the container did not have to be a rectangular box, how might the problem change?

h. What problems might occur with some of the containers discussed in part (g)?

1. **a.** If you were given a table of *x*- and *y*-values for a function, how could you tell if the function was a model of direct variation or inverse variation?

   b. If you were given a graph of points, how could you tell if the graph was a model of direct variation or inverse variation?

   c. Give an example of a real-life function that is a model of direct variation and an example of a real-life function that is a model of inverse variation.

2. **Open-ended Problem** The table at the right gives the grades for the last quiz in an Algebra class. To find the mean of the grades, Laura found the sum of the grades (33) and divided by the number of grades (14). She got an average of 8.25. The correct average is 9. What did she do wrong?

Quiz grade	Frequency
10	6
9	4
8	3
6	1

3. **Open-ended Problem** Compare rational expressions to polynomials. Are there rational expressions that are also polynomials? If so, give examples.

4. **Open-ended Problem** Write three different rational expressions that simplify to the same expression.

5. **Group Activity** Simplify the expression $\dfrac{x^2 - 6x}{x^2 - 2x - 24}$. Verify that your answer is equivalent to the original expression. Using several values of *x*, make a table of values for the expression $\dfrac{x^2 - 6x}{x^2 - 2x - 24}$ and for your simplified expression. Set each expression equal to *y* and graph the resulting rational equations on a graphing calculator. Discuss your findings.

6. **Performance Task** Is $\dfrac{x^2 + 3x + 2}{x + 1} = \dfrac{(x + 2)(x + 1)}{x + 1} = x + 2$ true for all values of *x*? Explain why or why not.

7. **Open-ended Problem** Compare simplifying rational expressions with simplifying rational numbers. Use examples in your comparison.

Chapter 12 ·····································

1. **Group Activity** Write an algorithm for finding a path through a maze. Draw a diagram of a maze of your own design. Trade your maze with another group. Use your algorithm to solve the maze from the other group. Write an evaluation describing how well your algorithm worked.

2. **Open-ended Problem** When you program a VCR to tape a movie, you are using an algorithm. Describe the algorithm.

3. **Project** During homecoming week at Pierce High School, an honorary court consisting of four female students and four male students is chosen. The following method has traditionally been used to select this court.

   On Monday of homecoming week, one female and one male representative are chosen from each homeroom. From these representatives, all the high school students will choose the court. On Wednesday, each student receives a ballot listing all the homeroom representatives. The student circles one female name and one male name as their choices. The four female students with the most votes and the four males with the most votes are on the honorary court.

   Design another way to organize the selection of the court. Compare your method to the method above. Write a letter to the student council trying to convince them that your method is better. Consider the fairness of both methods in your letter.

4. **Performance Task** Plan a travel wardrobe of several coordinated outfits. Include at least 2 skirts or pairs of slacks, 3 sweaters or jackets, and 4 blouses or shirts in your wardrobe. Draw or cut out pictures from magazines of your clothing pieces. Using a tree diagram, make a poster display of the different outfits that can be created from your wardrobe. Can you think of a way to use permutations or combinations to count the number of possible outfits?

5. **a.** What is the difference between a permutation problem and a combination problem?

   b. Describe two situations that could be counted using permutations.

   c. Describe two situations that could be counted using combinations.

---

6. **Performance Task** Mom, Dad, Laura, and Tyler are going to a play performed by a regional theatre company. They have tickets for four seats in the sixth row.

   a. How many different ways can the family sit together for the play? List all the possible seating arrangements. Use the permutation formula to confirm that you have the correct number of arrangements.

   b. Due to occasional misbehavior, it is best if Laura and Tyler do not sit next to each other. On the list from part (a), cross out any undesirable seating arrangements. How many seating arrangements are now available to the family? Use the permutation formula and subtraction or division to confirm that you have the correct number of seating arrangements.

Assessment Book, ALGEBRA 1: EXPLORATIONS AND APPLICATIONS